CULTURE, EVOLUTION, AND MENOPAUSE

CULTURE, EVOLUTION, AND MENOPAUSE

The Deceptive Dichotomy Between Nature and Nurture

Hippokratis Kiaris

Algora Publishing
New York

Names: Kiaris, Hippokratis, author.
Title: Culture, evolution, and menopause: The deceptive dichotomy between nature and nurture / Hippokratis Kiaris.

Description: Humans have always tried to overcome the burden of biological limits by developing science, technology—and communities. Canceling the link between biological or fitness and differences in reproductive success, humans have defeated the logic that guides evolution. But breaking the link between biology and fate results in an existentialist crisis.

Library of Congress Control Number: 2024946910

Printed in the United States

To Ioulia

Table of Contents

Introduction

The discussion on the borders between culture and biology, or nature and nurture as it is commonly called, is a very popular one. Nature refers to everything that relates to our biological substance as human beings, while culture refers to everything that characterizes us and is the intentional product of our intellect. Like our biological substance that is transmitted throughout the generations by our genes and evolves, culture is also transmitted throughout the generations and evolves. One influences the other, so they co-evolve as a unit, raising constantly the question of whether they are truly distinct, and whether one can be perceived without the other. We are always surprised when we find out that other animals, especially primates, do something that we do too, something that so far we had been attributing to our culture. This by itself implies that this something must be part of our nature, expanding the sphere of nature versus that of culture.

At the same time, we realize that the ways we live our lives today are determined mostly by the culture we live in. The two frequently appear to oppose each other, and each usually antagonizes what the other instructs, with biology usually being the limiting factor of the human condition; we constantly try to overcome it with our culture. Over the years we see an intentional attempt to increase the prominence of culture, and today we have reached a point where our lives are guided almost exclusively by the culture we have established and by its products in science and technology.

By examining common themes from the onset of human culture very early in our evolution, and by focusing mostly on the culture of the West, we will attempt to predict the future of mankind and identify the limits.

We will then examine our new condition as cultural constructs that carry a biological component that we try, intentionally and not, to minimize.

My children Frosini and Harris have been an undeniable source of inspiration in writing this book as, consciously and not, they placed in perspective things that would otherwise have remained beyond my notice. Also essential for the development of these ideas were the various discussions with Honors students of the University of South Carolina who took classes that I teach and graciously shared their concerns about the future that they will live in. Finally, I wish to thank Andrea Secara from Algora Publishing for her meticulous editing and feedback that clarified thoughts and ideas and smoothed the original draft.

1. Randomness in biology and the meaning of universals

Randomness

This initial chapter could very well have been positioned as the last one, but after several switches back and forth, I decided to place it at the very beginning. This way I believe the reader will become more familiar with the central idea of the book as it unfolds and will be able to see where this all goes, if it really does go.

Evolution as we understand it since Darwin's seminal work and the contributions of numerous biologists after him that interpreted it, enriched it, and aligned it with the continuous discoveries in biology, relies heavily on randomness: on the chances for gametes to come together and produce certain phenotypes, on the chances of mutations happening, and on the chances that at a given time a combination of certain environmental factors may exist to determine which phenotypes, and therefore genotypes, will be favored and selected. Therefore, randomness is inherent in biology to the extent that everything that is related to its subject matter can be understood and will make sense only by considering its probabilities to do so. In that sense, when we render a statement that conditions A will result to an output of B, what we really mean is that conditions A will most likely result in the output B. But we will get into that later. In the meantime, we need to elaborate more on what "being random" means for biology, and if this is different, and in which way, from the meaning that randomness possesses in the other natural sciences. This will enable us to appreciate what we examine from the biological perspective and attribute its value to what has happened already.

To start with, we need to appreciate that the concept of randomness is very distinctive in biology. This distinction defines the thin line that separates biology from philosophical inquisition, and it is responsible for fundamentally distinct approaches towards some of the essential questions that concern both disciplines. Furthermore, it defines another line of thinking, the one that distinguishes biology from other natural sciences such as physics and chemistry and determines the limits of each in describing the natural world. This is the topic that concerned from early on, Monod (1974) in *Chance and Necessity—An essay on the Natural Philosophy of Modern Biology* when the molecular basis of biological phenomena started becoming tangible for both the experimentalists and the theorists, and it continuously concerned equally the moral philosophers, the philosophers of science, the evolutionary biologists, and the cognitive scientists, when addressing questions about free will, determinism, or teleology.

Let's see what randomness is. According to Merriam Webster Dictionary, *randomness reflects the quality or state of being or seeming random (as in lacking or seeming to lack a definite plan, purpose, or pattern)*. The use of the term *seeming* adds a layer of extra complexity in our interpretation because it implies that the phenomenon in question may itself be subjective. Nonetheless, randomness reflects our inability to render predictions and determine future outcomes. Thus, randomness must be subjective and possess a time-dependent constraint. What we see as random now may not seem random to someone else at a different time. The eclipses of the sun or the moon in the far past reflected expressions of the gods' power and will, until the point at which our ancestors, in different civilizations, started predicting them by using the science they had developed. The weather of the coming days offers another example; in the past it was seen as random, but today, the sophisticated equipment we have invented, and the theories we have developed that allow us to use them, make prediction possible. In theory, having at hand tremendous calculating power and abundant data, we could accurately predict the weather 20, 30, and likely in the near future even 300 days out, as well. This is because what the weather is going to be depends on a number of factors that occur today and can be recorded, and it represents a direct function of these factors. It is not a state that will occur independently of these factors.

Similar to this is the outcome of flipping a coin, the prototypical stochastic experiment in both of the two possible outcomes, heads or tails, possess equal chances to occur at each time we repeat the experiment. Yet the flip of a coin follows certain mathematical equations borrowed from the field of mechanics. If we had a supercomputer available that could take into account and handle these factors, together with all the parameters influencing the coin flip, we could indeed predict the outcome of every trial. Assuming that the force of our finger, the way we hold the coin when we push it, the angle, the composition and pressure of the air, as well as other environmental factors that can influence the coin's rotation, were identical, the outcome would be always the same.

Even more complex systems that relate to life itself remain, at their essence, physical and chemical systems that follow the same principles that apply to the physical world and could have been predicted, provided that we had tremendous calculating power and data available. By considering the environment as a factor in this process and the chemical nature of the events we regard as mutations when they occur at the level of DNA, we could also conclude that even evolution might have been predictable. To that end, evolution is the outcome of a series of events initiated by certain chemical reactions and the mutations that generate genetic variations which are subjected to selection—to persist or perish. This should happen in a manner that is deterministic because all the associated phenomena are physical phenomena that are subjected to the physical laws of nature. In that case, of evolution, the apparent randomness may only reflect our limitations in acquiring data, analyzing them, and then coming up with models to explain them and to put them into the wider context. Irrespective of whether we can currently model these events, they are events that can be modelled in a manner such that progress will eventually render the models more and more accurate over time. This is a linear process by which models can only become better. Otherwise, they would not replace the previous ones.

Another example that supports the view that the limitations in our current experimental approach and calculating power do not compromise the definiteness of certain processes in the natural world may be borrowed from Xeno's dichotomy paradox. According to Xeno, in order to reach an end point we constantly need to move halfway towards this, for indefinite times (Aristotle, Physics VI:9, 239b10). Although the division of the distances is indefinite, the position of the point is definite and precise.

The calculation of the exact value of π provides another example of something that is tangible and exact, corresponding to the ratio between the circle's periphery and its diameter, but impossible to define with precise accuracy. Thus, we can conclude that definiteness exists, irrespective of the indefinite number of the processes in between. We can have an indefinite number of models, all describing with higher accuracy a phenomenon that is rather definite and precise. We can also have a definite phenomenon albeit we cannot describe it with absolute accuracy. That something will rely on approximation for its description does not compromise its exact nature.

The predominant perception that currently underlies the biological and the other natural sciences is to deconstruct each system into its elements and to understand it by applying laws from a discipline that possesses a subject matter that operates at a scale below the one that it originates from. For example, we try to understand sociology and psychology by infusing them with principles that are applicable to biology, and biology by applying chemical principles, and to understand chemistry by viewing it as a subject of physics. By doing so, we can argue that we identify causality and mechanisms. In brain function, for example, we strive to reduce changes in a person's mood and their behavioral responses to chemical balances and neuronal alterations, in such a way that we become increasingly capable in terms of both prediction and effective intervention due to the increasing technological progress we constantly achieve over time.

From that perspective, randomness reflects the breadth and depth of our limitations in modeling the system we examine, which in turn is related to the technology we have available, and in the theory that we have developed for backing it. This can be true in the natural sciences which rely on the laws of physics and chemistry and for which the probabilistic element is only related to the complexity of the system and of our current ability to dissect it. These sciences were developed and perceived as such, in a manner according to which the underlying phenomena reflect states that depend exclusively on the status of the preceding states. From the opposite direction, any given state can be followed by a specific given state.

Is this the case, however, for biology? Instead of trying to answer it, let's assume for a moment that this is the case and let's then see where this will eventually lead us. The premise is that evolution, indeed, follows steps that are deterministic in their essence, but we view them as random just because we don't have means of not seeing them as such. In other words, the mutations that differentiate our genome from that of our evolutionary ancestors by 1% were just chemical reactions in our nucleotide sequences that would have happened deterministically, albeit we do not—and most likely we will never—have computers and data that make us capable of performing the corresponding calculations and predict them. This represents the basis of determinism in sciences, according to which physical phenomena follow certain principles, while the probabilistic factor only reflects our inability to capture them appropriately. Irrespective of whether we will ever be able to do so, the corresponding processes remain deterministic, and the appearance of chance arises solely from our limitations in available methodologies and intellectual capabilities. From this perspective, randomness exists is only a construct of our mind, a construct that reflects our limits.

If we accept this line of thinking, we will also have to accept something else, though: namely, that the evolution of all species, including our species as humans, was somehow predetermined since the time zero of the natural world, from when the physical principles that guided it were set. Even though we are incapable of describing the physical events that occur in their entirety, such events are definite, they are describable, and would follow a unique set of physical principles that would repeat themselves when identical conditions exist. It does not really matter if we probably will not ever be able to develop appropriate technology to calculate it and predict it. The sequence of events should be the same since the previous state of things was identical. Thus, to repeat, evolution should conclude again to *H. sapiens* because all the individual events would happen again, and in the same order. This equates to the argument that the conclusion of the evolutionary process that led to *H. sapiens* was already determined when the clock of evolution was set, a few billion years ago when life started.

This sounds rather unlikely, though. It represents an idea that biologists will hardly find acceptable. Maybe it represents a limitation that we will never be able to circumvent, since it relates to the essence of our existence as species with given capacities and abilities, but it defines the world as we understand it and the limits of our potential. Like in Plato's cave allegory, our senses, and by extrapolation our technologies, will never allow us to see the world beyond the constraints of their limitations, but by appreciating this we may develop new understandings by which such limits will be integrated into our thinking and the perceptions of reality. It becomes admittedly controversial that our science, based as it is on the premise of determinism, fulfils itself with the acknowledgement that it operates in a stochastic realm.

Coming back to the original question of what randomness is, we may classify it into two different categories: the randomness that reflects our inability to render predictions because of unknown variables, or variables that cannot be integrated into our model (because of their magnitude and their complexity); and the randomness that results from the fact that a given state does not reflect the previous one. In the first case, randomness is just a construct of our mind, while in the second case randomness does exist as a component of our reality. The difference between the two is not clear and can yield different values at different times and in different contexts. Perhaps an easier way to distinguish them, instead of defining them, is to consider their consequences: In the case of methodological randomness, no matter how complex and unpredictable a system is, the result should always be the same, provided that identical conditions were established at the very beginning. However, in the case of genuine randomness, the result could be different even if such conditions were identical. Naturally, determining what 'identical conditions' means and whether such conditions can truly exist is another open question that may affect our conclusions.

The difference between the two is not clear and can give different value at different times and in different contexts. Perhaps we can distinguish them more easily than actually defining them, if we try looking at their consequences: In the case of methodological randomness, no matter how complex and unpredictable a system is, the result should always be the same provided that identical conditions were provided at the beginning. By contrast, in the case of genuine randomness, the result could turn out different even if the conditions were identical. Naturally, we have to ask ourselves how we can be sure we know what "identical conditions" even means, and whether such conditions can really exist. The answers to those questions, of course, may direct our conclusions in certain directions.

In the realm of natural sciences, at least at the level of the macrocosm that we operate in and perceive through our experiences, randomness refers to the first case, in which chance is only applicable due to the complexity of the factors that constitute the corresponding phenomena and our inability to integrate all available (including the plausible) data, in order to render predictions with accuracy. In biology, however, randomness extends to a different level. This is imperative and a prerequisite in order to rule out the possibility that evolution is predetermined, and that the appearance of our species is predestined. We may always maintain in the back of our heads, and entertain the possibility, that this distinction is artificial and that randomness, beyond the technological limitations that relate to the complexity of the corresponding systems, does really exist. In that case, such randomness is inherent to us as humans and always will be, for as long as we see biology as a science that bears a distinct subject.

Universals and historicity

How this applies to biology—beyond ruling it out because of its detrimental implications in our existence—is hard to perceive. If however, we lean towards accepting the occurrence of the second type of randomness, that a given biological state does not necessarily reflect a function of a previous state, and if we accept the fact that two outcomes could have happened out of which only a single one did, this may provide explanations and define the subject matter of biology.

It raises another important question, though, that also makes biology deviate substantially from the other natural sciences. The question is related to the operation of universals. In the latter natural sciences, such as physics and chemistry, universals do exist. They occur by a manner that is independent of time and space, and this is what makes us see them as universals. They are not conditional, occurring under specific circumstances only. The apple will always fall at a given speed and direction, because of the laws that govern gravitational forces, that in turn exist and operate irrespective of whether there is an apple that falls. The apple makes them visible to us and is subjected to them as soon as it exists, but the laws that govern its motion pre-exist the falling apple.

This implies a rather distinctive point for biology, one where the evolution of *H. sapiens* cannot be perceived without its outcome, the humans. We cannot refer to the evolution of *H. sapiens* unless we do have the *H. sapiens*. More globally, evolution cannot be described unless it is tightly linked to species and processes that can deviate from each other, by being subjected to different strategies. This is because evolution does not really represent a global scheme of rules and mechanisms but rather a retrospective analysis of what has already happened, presented in a manner that makes sense to us when we try to describe and explain them. The historical factor is intrinsic to evolution and by extrapolation to biology as well (Kiaris, 2024). This is what makes evolution unpredictable to us: When it happens, we can identify and describe the underlying principles and the corresponding dynamics, but before it happens, we cannot really tell how it is going to proceed. From that perspective, the subject matter of biology appears more relevant to history than to physics. Both history and biology can explain the past but cannot really predict the future.

The subject matter of biology exists because of its outcome, compromising the universality of the principle. One could argue that the genetic code represents the only probably true universal that exists in biology. Yet, even this as a principle is illusory because it has emerged as a universal just on the grounds of its historicity. The genetic code had evolved very early, at biological time zero, and after that, and actually because of that, it persisted. This really provides unequivocal evidence of the fact that all living organisms originated from the same ancestor, but this is not sufficient to attribute universality with the same rigor that we can attribute to the universals that exist in the realm of physics. There are no solid grounds that would suggest that a quartet or a doublet code could not have happened. If they did and persisted, then the quartet or a doublet code would have been the universal. Furthermore, the co-existence of organisms with the triplet code and a theoretical doublet code is also plausible because these different codes are not mutually exclusive.

Even more strikingly, we are close to having the technology to manipulate the genetic code. The genetic code is the algorithm that makes the correspondence between three specific bases of DNA (or mRNA, to be technically accurate) and one or more amino acids that are used to compile the proteins. This happens by engaging specific RNA molecules, the "transfer RNAs" or "tRNAs." On one side they recognize the three bases of mRNA and on their other side they carry the amino acids. If we wanted to, we would soon be able to a construct tRNAs in the lab that might, for example, carry different amino acids that are not used in proteins, or to recognize doublets in the mRNA, to do so. In principle, no one rules out the possibility that even bio-constructs that are based on sulfur instead of carbon can be built and self-propagate by a manner that resembles actual cells. Of course, the question in that case would be whether those bio-constructs would continue to fall into the subject matter of biology or would define a new biology.

Biology is more analogous to a construct than a science: something that over time changed to entertain strategies. These considerations emphasize the role that the historicity, or the lineage, has in biology as being what determines the significance of universality. They suggest that universals in biology are indeed universals for historical reasons and not because they reflect principles that persist across space and time. The principles of biology are compatible with other principles that could have happened but did not do so. Yet they can always happen, expanding the definition of what universal means (Kiaris, 2024).

Such relativism is foreign to the other natural sciences, where universal principles are universal and incompatible with other principles that may co-exist with them in the field. An apple that falls down is incompatible with an apple that goes up in the same world. In biology, two organisms that co-exist but have different genetic codes are conceivable. The genetic offspring of a single mother or father would have been inconceivable and against the biological universals until a few years ago. Today they are feasible, and they did not happen as yet only because of our conscious decision not to implement them.

A caveat in these arguments comes from the domain of physics, where quantum mechanics recognizes true chance functions that compromise the perceived universality and determinism that govern classical physics. "*In orthodox approaches to quantum mechanics, some measurements of a system in a given state will not yield a result that represents a definite feature of that prior state*" (*Albert 1992*). In that case, we could see biology and evolution as the sum of quantum alternatives and decide that the compromise of deterministic thinking does not apply to biology only but to other natural sciences. Such arguments only raise the impact of randomness as a state of the natural world that is wider than what we may think.

The two paradoxes

An implication of these considerations is the emergence of a paradoxical state, no matter how randomness is defined. According to the first possibility—that randomness is only methodological—we would have to assume that the same events that have happened already since biological or cosmological time zero would happen again. In other words, our evolution and our emergence as a species was predetermined and unavoidable. In that case, we would need to assume the occurrence of a teleology in biology, a notion that is abruptly rejected in natural sciences. In that case we would be required to accept biological hierarchies. In such hierarchies *H. sapiens* possesses a unique and superior position on the ladder, which again instructs us to revisit our role in the universe and redefine our priorities as species. If *H. sapiens* represents the teleological conclusion of a process that started about 4 billion years ago, when life appeared on Earth, then we should treat ourselves as an ultimate biological entity that needs to identify the causes of its existence. In that case the evolution of culture, a unique feature of our species, acquires distinct significance as it marks the distinct point of the teleology we are engaged into.

Another consequence, equally paradoxical, emerges if we assume that randomness is not only methodological but genuine as well. Despite the fact that the macrocosm within which we operate does not witness it, it is very likely that such randomness, operating on a different scale, in microcosms for instance, would also affect the macrocosm in ways that we cannot appreciate. In that case we would also need to accept the randomness as a component of the natural world and that all our efforts to eliminate it are, in their essence, futile and destined to fail. In that case, the whole scientific enterprise we have developed suffers from severe limitations because it is tasked to explain phenomena that can never be explainable since they bear randomness.

In both cases, a plausible resolution would be to re-invigorate an anthropocentric perception of the world, due either to the teleological considerations of the existence of non-randomness or to the acceptance that genuine randomness does exist and underlines our relativistic view of the world in a manner that is consistent with our limitations and capabilities. As we will discuss in subsequent chapters, the whole goal of humanity throughout its history, as a biological and cultural entity, has always been to generate environments in which these limitations have the smallest possible impact. In other words, the aim has been to defeat randomness. In the case where randomness is methodological, we try to do so by perfecting our technologies and sciences to predict more accurately what will happen, and then we interfere with the outcomes. In the case where randomness is genuine, we aim to develop an environment that, in its broadest sense, becomes resistant to the effects of events and factors that can be perceived as random.

Whatever the case is, and irrespective of whether and how such paradoxes can be resolved, we can move forward by taking into account that, at least in biology, historicity and randomness possess a unique position that governs how the phenomena that relate to its subject matter unfold. On the other side, however, we can also recognize patterns that emerge and become apparent at different scales, and in contexts that appear irrelevant, pointing to a common theme that governs similar processes and recalling notions that relate to universality. In evolution, for example, we see similarities in forms of organs and tissues that have quite different and indeed distal if not irrelevant origin, such as the eyes of insects and mammals or the "flowers" of plants or yeasts. In behavior also we frequently see underlying patterns that appear similar albeit evolutionarily irrelevant. One example would be the evolution of monogamous behavior that, even within some highly related species of the same genus, have evolved independently at least more than once (Leslie et al., 2010). Similar forms create patterns that can evolve in response to similar conditions, to resolve analogous needs, that can point to universality.

At the level of human cultures, we can also recognize analogies between the lifespan of civilizations and the different stages of personality and behavioral growth in children (Kiaris, 2023). Different civilizations, such as the Greek-Western and the Chinese, although they did not come into contact, at least at a considerable degree and early enough—sufficient to explain their co-evolution and reciprocal influence—have passed through different stages that resemble the stages of development of children, from early on when they are adolescents to adulthood (Kiaris, 2023).

In all cases, of course, we cannot formally rule out the possibility that the similarities we recognize—our tendency to identify universal principles—is just an illusion of our cognitive abilities that strive to detect order within chaos. Similarity, divergence and difference are relative terms that acquire only the meaning we can attribute to them. The extended nuclear Western family and the band of genetic relatives in tribal organization represent a similar mode of social organization because we want to see them as such. With the same accuracy that we can outline the similarities, we can also outline the differences.

Strictly speaking, while universals may not exist in biology in the way we think they do in other natural sciences, such as physics, certain manifestations remind us of universal principles that allow us to study the corresponding processes and generalize when we try to identify patterns and common themes (Kiaris, 2024). It is up to us, the observers, whether we emphasize what is similar and what is not, how stringently we define a pattern when we see it, and how tightly we will adhere to the corresponding forms. Nonetheless, the recognition of such patterns and norms, at the scale of human populations, defines cultural trends that emerged very early in the evolution of our species, and since then they have become prominent, guiding us until today.

The evolution of human culture should be seen from that perspective, as a force that uniquely developed in our species and possesses a transcendental power that is governed by its own dynamics. Within this process, culture acquired increasing dominance in peoples' lives that reduced the power that mere biology had so far held ever since *H. sapiens* evolved. This was done especially by the culture of caring we have developed which progressively renders resources available to benefit an increasing number of individuals, at least as regards their essential biological needs of food and shelter. People now come into being not only as biological entities in a given physical environment that is less hostile and challenging than it was ever before but also as the carriers of a cultural heritage that pre-exists them and shapes them. By defeating biology, to the extent that this is feasible, through science and technology, people come to appreciate an existential crisis as they need to determine for themselves whether borders between nature and nurture do truly exist. As such, and to the degree that culture represents an inherent characteristic and an unavoidable product of our biological substance, we come to the realization that such borders are rather artificial, offering a sense of liberation that allows us to revisit our goals and determine our purpose from a perspective of reducing the biological challenges we constantly face. Within this new context, biological randomness acquires a more prominent role, as a paradoxical by-product of the increasing cultural stringency we have immersed ourselves in.

2. Forces of evolution

Extinctions

Since the inception of biological time, when the first forms of prokaryotic life originally appeared on Earth around 4 billion years ago, species continuously appear and disappear through a never-ending process that we describe as biological evolution. Some species, or genera, have seen fewer changes than others over time, such as the archaea. They are prokaryotes that originally evolved about 2.5 billion years ago, and as their Greek-derived name indicates, they are ancient life forms.

Other species had shorter lifespans. The ancestors of the woolly mammoth that went extinct around 10,000 BC originated in Africa around 3 million years ago and then spread into the Northern Hemisphere. The exact reasons for the extinction of the mammoths are not clear, but they are most likely associated with the climate change and the melting of ice that changed their conditions, as well as hunting by humans that by that time had already spread. The dodo bird is another example of a species that went extinct recently. Dodo birds are large, non-flying birds that lived exclusively in Mauritius, evolving from ancestors that migrated there about 8 million years ago. They went extinct in the late 17th century, after Dutch sailors claimed the islands and used dodos as food, and introduced domesticated animals such as cats and pigs that preyed on dodos. The Rocky Mountain locust, a grasshopper living in the Western United States and Canada and a major pest for the pioneers, disappeared in the 19th century after their natural habitat was transformed into farmland.

When conditions change abruptly, species disappear at higher rates. So far, life on Earth has experienced 5 major extinction events (Greshko et al., 2019). The most recent one happened 66 million years ago at the end of the Cretaceous Period and is attributed to an asteroid that hit Earth. Some 199 million years ago, in the late Triassic, massive volcanic eruptions in what is today the Atlantic Ocean caused another massive extinction. At the end of the Permian, 252 million years ago, volcanic eruptions and global warming resulted in the largest extinction event in history. In the late Devonian, 378 million years ago, the extinction of many marine species occurred, but the causes of this event still remain vague. In the Late Ordovician, 447 million years ago, the earliest extinction event occurred, which is related to global cooling and glaciation.

Today it is thought that the sixth extinction is underway. This extinction, contrary to the previous ones that are attributed to conventionally viewed natural causes, is associated with human activity. It is often called the biodiversity crisis and is associated with the effects that our species has had and continues to have on the environment (Rafferty, 2023). It is estimated that since the arrival of *H. sapiens*, species have been disappearing at a rate that exceeds by 100 to 1,000 times the rate at which they did so in the past, before humans occupied the planet (Wilson, 2023). This is related to various anthropogenic activities, both direct and indirect. For example, hunting, since the earliest stages of human history, has contributed to the disappearance of mammoths and other species as well. The effects were more severe in subsequent periods when humans started cultivating the land and thus displacing the endemic species that have now lost their natural habitats. Even more detrimental are the modern consequences of industrialization and contemporary global warming, which have gravely accelerated rates of extinction by drastically changing ecosystems, both intentionally and unintentionally, and both directly and indirectly. After acknowledging this new threat to what we perceive as our natural environment, systematic efforts are now ongoing, funded by both private and government agencies, to record biodiversity and rescue species that are under threat. Furthermore, regulatory activities in the form of laws and regulations are continuously being implemented, and continuously also being criticized, to minimize the effects that human activities have on the environment.

From a purely biological perspective, although the rate of species extinction is far greater than the rate of new species that emerged in the Anthropocene (the period that coincides with the presence of *H. sapiens* and his effects on the ecosystem), indeed several new species have appeared because of humans (Rejón, 2021). The domestication of animals and plants represents direct evidence of how humans have promoted the development of new species for their benefit. In that case we can interpret this as a form of symbiosis between us and the domesticated animals, because we need them as food or for agricultural and other activities, while they are not likely able to survive by themselves in nature. This was done purposely as a choice because a need had to be met, and this need was the food and resources for the increasing human population.

In parallel, as the environment changed, new ecological niches appeared in cities, industries, mining fields, oil platforms, and waste disposal fields, all of which created new conditions that could host new species with different requirements from their ancestors. Through this process new plants, birds, mammals, insects, fishes or prokaryotes evolved and occupied these niches, species that had been adapted for life in these environments and that are quite distinct from those that existed previously. By occupying these niches they faced reduced competition from other species.

Drivers of evolution

Textbook knowledge in biology recognizes that evolution proceeds by mutations, genetic drift, natural selection, and gene flow. Mutations are changes in the DNA that occur stochastically and result in the development of new forms of genes, or alleles as the versions of the genes that carry these specific mutations are called, that produce proteins— albeit not all genes correspond to proteins—with slightly or drastically different properties than the original ones. If the proteins that these new genes encode for, are associated with functions that are more efficient than the original ones, then natural selection will favor the individuals that bear them and will persist, because they will increase their fitness.

It becomes obvious that it can be determined that a function is performed more efficiently only within the specific environment the individuals live in. For example, a gene that offers the ability to live at a high concentration of oil will be useful only for populations that live around the oil platforms. Elsewhere it will be either useless or it can even be harmful. To appreciate this from a wider angle, we can recognize the utility of lungs only for organisms that live on the land or in the air, and not in the sea where gills are needed for oxygenation.

Fitness is a genetics term that reflects reproductive success and indicates the adaptation of individuals. It is used as a measurement to determine how changes in specific genes influence the ability of individuals to adapt to a given environment and under certain conditions. Better fitness means more offspring that are capable of reproducing; it represents the complex sum of different variables. This can be because, for example, a genetic change makes the individual faster or taller, which in turn makes them able to avoid predators or obtain food more effectively than one that is slower or shorter, respectively. Similarly, a mutation that increases the activity of an enzyme that is needed for food digestion will increase fitness because the conversion of food into energy will be faster and more efficient.

Nevertheless, there is always some cost. Running fast is good, but it also requires energy. Therefore, the individual that bears this mutation needs to be able to back it up with access to better food, energy-wise. Otherwise, it is like having a very fast car but not having good quality gas to fuel it. Or having a fast car but not having roads to drive it at high speeds. A hypothetical mutation that occurs in proteins that are involved in the maturation of the oocytes will not necessarily increase their fitness if it stimulates the production of more oocytes, because the offspring need to be fed and taken care of. There is a certain amount of milk that a lactating female has the capacity to produce, which limits the number of offspring she can nurse. Thus, such mutations, in order to be selected, frequently need to be corroborated by other mutations in other genes that can support them, and they in turn support the other ones as well by depending on them.

Humans are not the fastest, strongest, or most reproductively capable animals. Actually, they rank pretty low in each of these categories when compared to others. Yet they are the smartest, which gives them the ability to circumvent these limitations by being innovative. They devise strategies and invent tools that allow them to overcome their weaknesses and turn them into benefits. More importantly, they are social and live in groups by developing structures in which they exhibit complementarity in their needs and synergy in their strengths.

A direct outcome of this is the collective intelligence they have developed and constantly rely on, since the inception of historical time. It makes them able to start their lives on an existing background that was set by their forebears. They don't have to start their lives from scratch, reinventing everything from the beginning. They start from an established foundation that was provided by their ancestors, and to their ancestors by their ancestors. All other animals must start their lives with the same background as their predecessors, and eventually they live their lives in similar ways to that of their ancestors. If this way of life becomes different, it is only different because the context has changed and not because of their effort. Humans start our lives from a standpoint that was established consciously by our predecessors, going all the way back to the establishment of our culture and of our civilization that shape the perspectives from which our lives are viewed and lived. Marginal exceptions are provided by animals such as the beavers, for example, that terraform through dam building. This, in turn, intentionally changes the environment in which the next generation may live. It is obvious, though, that this process does not occur at a scale that compares to what humans do.

Such advancements were truly revolutionary for the human species. These advancements became possible due to mutations in genes regulating the activity of our brain cells and offered us capabilities that our evolutionary ancestors did not have—or if they did, they possessed them at minimal level. These are matters that we will revisit later, when we discuss the impact of our sociality on our cultural evolution.

Benefit, of course, is not simply due to the emergence or the improvement of a specific character that contributes to enabling us to perform a task in a way that is more efficient than how it was done before. Opposable thumbs, for example, are a character that several primates have, including humans, and it offers the ability to grasp objects firmly. In humans, however, it was further refined and used, in parallel with the development of more advanced brains, for the making and use of sophisticated tools. Benefit can also relate to a character that, in the changing environment, makes the individuals better adapted. For example, the distribution of body fat in a manner that assists in insulation in colder climates when the weather becomes colder, or the loss of the ability to fly in the absence of predators in birds inhabiting certain environments.

The pattern has always been the same: new characters have emerged or been lost through mutations that, when they were beneficial, were selected and transmitted to the offspring that also bear them. Benefit was judged by the number of offspring that were also able to reproduce. Eventually, the accumulation of these characters became high enough to separate the individuals that had them from those that did not, which made them become a new species. In that case, a necessary factor should also be their reproductive isolation, or in other words, the inability of the individuals with the mutations to breed with those without them and produce sufficient fertile offspring. Otherwise, the mule (the hybrid of donkeys and horses that is used extensively in agriculture because of its strength and resilience to hardship) would constitute a species by itself—if mules were fertile, and if breeding between a male and female mule produced mules as well.

When the mutations happen, their impact on gene function and the organism is unknown. It can be beneficial, harmful, or neutral. The ones with a positive impact are selected, while the ones that are neutral and do not interfere with fitness can persist in the population, just because they are not purged out of it. In various cases, though, mutations that are harmful also persist in the population. They do so because their negative impact is masked when at least one allele is normal (or 'wild type,' as it is conventionally called). In these cases, both alleles have to bear the mutation, one inherited from the mother and the other from the father. They need to come together and coexist in the child in order for the disadvantaged phenotype to exhibit.

This does not happen frequently because the chances of both parents being heterozygous are small. Several disease-causing genes, such as the HEXA, for example, that is responsible for Tay-Sachs disease, a progressive neurodegenerative disorder with an incidence of 1 in 320,000 births in the U.S., belong to this category (Ramani et al., 2023). No apparent benefits are associated with the mutant forms of HEXA that are constantly generated by the functional allele, and they persist in the population because they are present at low frequency, which drastically reduces the chances of being inherited by both parents and thus causing the disease. Therefore, the non-functional alleles are maintained at low frequencies in the population for several generations. If, because of random events, the frequency of the nonfunctional alleles increases, the increased prevalence of the phenotype in the population will eventually result in strong negative selection that will be more prevalently manifested, or in other words, in its reduction in frequency over time, to levels that will not pose dangers for the stability of the population. Several genes that cause pathologies associated with family history persist by these mechanisms through which they maintain (or are in the process of reaching) an equilibrium in their population rates, and the frequency of these deleterious alleles remains low.

In other cases, the behavior of a gene is not positive or negative on its own, but its behavior depends on the status of the other. The mutant allele, albeit harmful in homozygosity, may become beneficial when it is present in heterozygosity, which provides a mechanism that explains its persistence in the population, especially in certain environments. The classical example of such a mutation is the one that is responsible for the sickle cell disease. When it exists in homozygosity, again because a mutant allele has been inherited from each parent, it is responsible for this devastating blood pathology. It encodes for a defective form of hemoglobin that carries the oxygen in our blood and distributes it to the peripheral tissues, changing the shape of the red blood cells and causing severe pain and death when unmanaged. Nevertheless, despite its detrimental consequences in homozygosity, the individuals that are heterozygous (bearing one mutant and one wild copy of the gene) not only are not sick, since the trait is recessive, but actually they exhibit resistance to malaria (Haldane, 1949; Piel et al., 2020). Thus, in regions where malaria is endemic, such as in Africa, bearing the mutant allele is beneficial. This clearly shows the impact that the specific environment has in classifying an allele as defective and explains why the allele persists in the population by benefitting heterozygous individuals.

The general scheme of how biological evolution proceeds is further complicated by additional factors that characterize the populations and their dynamics from a genetic perspective. A theoretical imperative in genetics, and in the transmission of genes from one generation to the next, is the equal possibility for mating between individuals and their sufficiently large numbers, which are necessary to maintain the allelic frequencies at stable levels. Furthermore, it assumes that each individual leaves the same number of offspring in the next generation. However, this is not always the case. Populations are not always large enough, and just by chance some individuals may leave more offspring than others. This is not because their fitness is higher but may be a consequence of random events. When the populations are large, then these effects become minimal, and the genetic frequencies are stable. However, when the populations are smaller, leaving more offspring can be impactful because the genes of these individuals will be over-represented in the next generation.

Genetic bottlenecks due to smaller population sizes can be extremely impactful and are a frequent cause of evolutionary changes, because they result in drastic changes in the allelic frequencies from one generation to another. Think of a big box that has 10,000 marbles, half of which are red and the other half are blue. Assume that you want to recreate the pool of the original 10,000 marbles, the equivalency of creating their next generation, by picking randomly some, and then by increasing their number while maintaining the ratio of the marbles that you have picked already. If, for example, you pick 1,000 marbles at the beginning, half of which are red and the other half are blue, then for each blue or red marble you add 9 new marbles that are blue or red, respectively. If you picked 490 blue and 510 red marbles initially, it would not make a big difference in the end because you would end up with approximately 49% blue and 51% red. However, if you picked only 2 at the beginning, in half of the cases you would end up having exclusively red or blue marbles among your 10,000 marbles. When the population of marbles is small, one color can easily disappear. This reflects the consequences of genetic drift and the founder effects associated with it, by which small population sizes alter considerably the frequency of the alleles in the population throughout the generations.

Founder effects have shaped human evolution. They can become common in certain communities where, due to cultural reasons, the effective size of the population is small (Diamond and Rotter, 1987). In these cases, the isolation is cultural and not geographical. For example, in the Amish communities of Pennsylvania, the Ellis-van Creveld (EVC) syndrome (a form of dwarfism) is more common than in the rest of the population because the defective forms of the EVC and EVC2 genes are represented at higher frequencies due to their inheritance from the original founders of this group (McKusick et al., 1964). These differences are sustained due to cultural factors. Similarly, a defective form of the USH1B gene is responsible for a congenital form of deafness that is common in Samaritans (Bonné-Tamir et al., 1997). The inheritance of different blood types in different human populations represents another example of genetic drift, according to which the specific genetic profile of a small founder population results in a population in which the allelic frequencies differ considerably from those of the parental population (Perry et al., 2012). Geographical reasons due to the genetic make-up of the original populations, and not cultural reasons, are responsible for these differences. In some cases, this can relate to differences in function and performance, or fitness, and in other cases it can be irrelevant, such as in the different blood types and their prevalence in different populations.

Historically, some populations may have been isolated for cultural reasons, even if no geographical barriers existed, resulting in changes in allelic frequencies that propagate over time. Geographic isolation also accounts for changes in allelic frequencies, and it was rather intense during the evolution of our species. When the out-of-Africa migration of *H. sapiens* occurred tens of thousands of years ago, founder effects contributed to the different characteristics that the different groups of people had when they colonized the different continents (Kiaris, 2022). This is the reason that today, all over the world in different populations, the allelic frequencies of several genes differ drastically, and some of them may be completely absent in various populations. All these polymorphisms can be found at variable frequencies in African populations, where all variation in the genetic makeup of humans originated. When, however, the out-of-Africa migration started 50,000–80,000 years ago at different instances and by different subgroups of people, the genetic pool that these migrants represented was not reflective of the original genetic pool of their ancestors. Therefore, differences in the new populations were established by founder effects and then propagated. We can identify examples of these differences today in the appearance and the various traits that these different populations maintain, such as the color of eyes or hair, the predisposition to various diseases, and behavioral and personality traits (Kiaris, 2022). In their new environments, several of these polymorphisms were subjected to selective pressure, and the ones that benefitted the individuals expanded further. Those that were particularly harmful were eliminated; others were selected, while the ones that were more or less neutral persisted as markers of our genetic history. The various genetic providers that perform genetic testing and inform us about our personal genetic heritage are based on the analysis of the frequency of exactly these markers. Our individual markers are compared to those of different populations, and conclusions regarding our genetic roots are inferred.

Gene flow represents a variation of this mechanism, by which the change in the allelic frequencies is mediated not by founder effects and drift, but by the actual migration of individuals, provided that the incoming and the recipient populations have different frequencies in the corresponding alleles. A modern example of this phenomenon is provided by the colonization of the Americas. The original migrants that colonized the continents thousands of years ago were the ancestors of the Native Americans; they established populations that were also guided by genetic drift and founder effects. Subsequently, from the 15th century and onwards, until today, populations from Europe, Africa, and Asia have been continuously migrating to the Americas, willingly or not, enriching the genetic pool and contributing to the genetic diversity seen today (Adhiyaman et al., 2017). Initially, these new incoming populations were reproductively isolated for cultural reasons that dictated different accessibility to mating partners, but over time, these barriers are progressively collapsing. Continuous genetic flow changes the genetic make-up of the American population and an analysis of the genetic variation can trace back individuals to their ancestors and the populations they came from.

Selection of genes and strategies

The examples discussed earlier focused on how frequencies of alleles can change, driving evolution. In these cases, selection was not the major driver of the changes in gene frequencies. Nevertheless, the typical driver of biological evolution is selection. It is tightly linked to the environment, which in its widest sense we can describe as context, and it may work both ways, to increase and to decrease genetic frequency. Ultimately selection translates to higher reproductive efficiency, to more offspring.

The new mutation that is selected may, when it occurs, enable individuals to perform a task more efficiently than the individuals that do not possess the new character, without necessarily being linked to a change in the context. For example, let's take the mutation we mentioned that makes its bearers run faster, so they can avoid predators. The ones that carry this new mutation will not be caught by their predators so their chances to survive will be higher, thus they will have higher chances to leave more offspring. In turn, these genes will have their frequency changed in the population since more individuals now will have the fast-running gene.

Alternatively, the new mutation may be linked to a change in the environment. We mentioned already the example of a hypothetical mutation that makes individuals regulate their lipid metabolism more efficiently so they can sustain a colder environment. Such a mutation can be linked to environmental conditions that reduce the temperature.

This unique and over-simplistic designation of mutations that depend on the environment becomes compromised and problematic when we think more deeply about these phenomena. For example, the fast-runner mutation becomes effective only when fast-running predators are present in the environment. Furthermore, a mutation that makes an individual run faster will unavoidably impact the predators that now will have lower prey access, which will subject them to selective pressure to run faster as well. Or alternatively, the predators must become more intelligent and hunt the fast-running prey by using their skills instead of running faster. Now the prey represents the new environment for the predator, and the roles can be reversed. In that case we can argue that these characters, in the predator and its prey, co-evolve.

A precise definition of what the environment is can be rather confusing and extends across the scales of biological organization. Even at the gene level, we may consider a specific allele as "self" but at the same time it can define context, the "otherness," for its homologue, to produce a phenotype. Let's revisit the sickle cell disease mutation. The mutant allele can be harmful in the context of homozygosity (or in the presence of another mutant allele), but in the context of heterozygosity, when its homologue is wild type, it can be beneficial by conferring protection against malaria (Haig, 2020).

Although all these may be simply viewed as just semantics, they have value because of the teleological significance they possess. In the second case, when the mutation is directly related to an environmental change, then it is adaptive and reflects the change in the context over time. The sickle cell disease mutant allele is beneficial only when malaria is present. Otherwise, it becomes irrelevant. Nonetheless, it prepares the population to deal with malaria if and when it appears. In the first case, however, when a mutation and therefore the change in the morphology occurs that improves the bearer but it is not associated with a change in the environmental conditions, then it results in the "betterness" of the individual, the improvement of their skills and capacities, making them more efficient than before.

Evolution changes the species constantly, in a process bearing a linearity that covers all ancestral species until the ones that are now present and beyond. Genes carry information that eventually translates into a character, materialized by the structure they encode for, that finds its biological value in the function it performs. This function can be more efficient or less efficient, depending on both the gene and the genetic information it possesses within the environment, and can also be both external and internal, such as the one that is limited by the gene's own properties or depends on the properties of other genes and their information, respectively. Thus, the "form" of what the gene translates into is intrinsic to this information and it is hard to acquire a simple value that is universal.

Forms in ancient Greek philosophy were extensively discussed by Plato and Aristotle to explain the world; they represented an idealized version of it, especially according to Plato, as it was experienced by us, the humans, through our senses and interpreted by our intellect. They had a vital difference, though. While for Plato the forms are idealized versions of reality that existed in the realm of forms independently of material objects, for Aristotle they were tightly associated with objects and could not be perceived or even exist without them (Ainsworth, 2020).

To draw an analogy with the forms of Plato and Aristotle, one can see the mutations that improve an existing character driving it into its "betterness" aligned with the Platonian forms, because Plato considered forms as the perfect version of reality, and reality as the approximation of the forms. Alternatively, the forms that offer adaptation into the changing environment, and thus cannot be viewed as perfect but just different, increasing the diversity, as Aristotelian. Nonetheless, Aristotle with his strict hierarchy and his directional approach towards the Unmoved Mover (to be discussed in the penultimate chapter) also recognized the overall betterness as a fact that guides diversity. Of course, he had not conceived evolutionary processes and the species for him represented just different, rigid versions of the real world.

From collectivism to individualism and back

In comparing the evolutionary dynamics of different types of organisms we will come to the realization of some interesting facts regarding how different genes operate in the context of the groups they exist within. In prokaryotes, as well as in the eukaryotes that reproduce asexually, the whole chromosome is a single unit that is transmitted to the next generation.

Thus, the benefit or defect that is conferred by a new mutation in a given gene will automatically apply to all other genes as well. Thus, they work together as a group, and as a group they are selected throughout the generations. Events of horizontal exchange of genetic information that increase their genetic diversity do occur, but their extent does not match the levels of diversity of organisms that use sexual reproduction, males and females, to reproduce. In sexually reproducing organisms, the maternal and parental chromosomes come together in different combinations in each offspring that is produced, in a process that systematically shuffles their genome in the offspring. Such functions render the whole genome of the asexually reproducing organisms as the unit which is selected and therefore predominates or disappears. In sexually reproducing organisms however, after each mating, new and unique combinations of genes, or genotypes, emerge, carrying genes that have already been selected individually in the past. A defect or a benefit in a single gene will affect the success of the other genes that coexist with it, but only temporarily for this generation and in this particular individual. In the next one, different gene combinations will emerge, and each individual gene will have the chance to be vetted and selected, again and again in multiple genetic environments. Thus, over time, the dynamics that characterize which alleles will persist go through several rounds of selection at which each gene is subjected to forces that are tried in different genomic contexts, or in the background of a multitude of gene combinations (Haig, 2020). The evolutionary trip of genes in this case of sexual reproduction is lonely and individual, while in asexual reproduction it is characterized by companionship, in groups of larger cohorts. The "otherness" was discovered when sexual reproduction evolved by defining what "self" is.

Asexual reproduction appeared earlier than sexual reproduction, about 1.5 billion years ago, as opposed to the almost 4 billion years ago that life started. Thus, the earlier stages of biological evolution were characterized by companionship and collectivism. The fate of each gene was intertwined with that of every other gene in the same chromosome and they were "doomed" to stay together in the future as well. And together they perished or triumphed. At subsequent stages, though, with the emergence of sexual reproduction, an individualistic dimension started being entertained, concomitantly with the comprehension of "self" and of the "other", of the "non-self". Initially, the success of each gene in the long run was determined not only by its own performance but by that of its companions as well, in the prokaryotes and the other asexually reproducing organisms. However, when sexual reproduction came into play, success was determined by the sole properties of the gene in question. Advantages and weaknesses conferred temporarily by its companion genes would eventually cancel out, offering a transient advantage or disadvantage, leaving as net loss or gain only the one that corresponds to the gene in question alone. The lower mutation rate in prokaryotes and in other unicellular organisms that reproduce asexually, notwithstanding the limitations in calculating it and the high variation among its members, warrants that the team that wins does not change (Drake, 1999).

However, that the genes, upon sexual reproduction, started behaving independently guided by their own good only, illustrates the prominence of "self"; but it is not always true because eventually the benefits to oneself in having appropriate companions began to have an impact. Complex dynamics started to emerge that increased the complexity in the interactions between the genes and establishing networks of interdependency. This applies both to the relationships between different alleles of the same gene, or a single genetic locus, as well as between different genes, at multiple genetic loci. Regarding the impact of "otherness" in a single locus, we have already mentioned the example of the different alleles in sickle cell disease where, when malaria is present, the wild type or normal allele is benefitted by the presence of the mutant allele that confers resistance to the disease. *Vice versa*, the mutant allele requires the wild type allele to confer heterozygosity, otherwise when being in homozygosity, the mutant allele would have its chances for survival of the individual carrier reduced drastically, which in turn would terminate the existence of the mutant allele as well.

This interdependency of the different genes in the genome results in the establishment of hierarchies in which some genes are more essential than others, in an essentiality that depends on the scale of biological organization. For example, the gene that encodes for DNA polymerase is essential for all cells' survival and proliferation, but a gene that encodes for a reproductive hormone only becomes essential at the organismic level, without compromising the survival of the cells that constitute the reproductive organs. It compromises in its absence only the survival of the species since without it, individuals will not be able to reproduce. By default, of primal position in this hierarchy is the gene encoding for the enzyme that is responsible for the duplication of DNA, the DNA polymerase. Without this gene, neither the gene for DNA polymerase nor any other gene in the genome could pass into the next generation. Another essential role, immediately below that one, is held by the gene encoding for RNA polymerase. This one is in charge of converting the information that is stored in the DNA, and then transferring it to the translational machinery of the cells, the ribosomes, that interpret the information to functions. These genes together account for just a few hundred out of the thousands of genes that exist in a eukaryotic cell, and the essentiality corresponds to the actual storage and the direct translation of the information into function. Essentiality can also be relevant to the level of the organism at which the genes do not influence the other genes directly, but only indirectly, by affecting the survival of the organism the cells of which carry the genome. Finally, there are stretches of DNA, typically classified as genes or not, that are just passengers: They don't carry any type of vital information and their absence does not compromise the function of the other genes or of the organism. These genes, however, receive, on an equal basis with all others, the benefits of the action of the other genes, although they do not contribute to the fitness of the cells, the organism, or the population. They are subjected to corrective actions that protect from mutations, and persist. The benefits in that case of existing within a group of genes, the genome, are distributed to all genes, disproportionately to their contribution in the others' survival. Their persistence is based on the tolerance of the other genes, the contributors, to allow the existence of the beneficiaries without developing active mechanisms to purge those parasitic entities out of the genome.

The complexity in the interactions between different genes is not limited only to the interactions between genes of different type, but also to genes of the same type but harboring different polymorphisms. This is the case of different loci, that tend to show correlated frequencies in different populations. In these cases, we can recognize a trend according to which when a certain allele has higher frequency in one population, other alleles would also have high frequency as well. For example, polymorphisms in the genes CTNNA2, OXTR, LINC00461, MTMR9, and WSCD2, all of which are associated with personality traits, show high correlation in their frequencies between Chinese and European ancestry individuals (Kiaris, 2022). When one allele exists at high frequency, another is also present at high frequency as well, and when the frequency of one is low, the other's is also low.

CTNNA2 has been associated with excitement seeking, OXTR with sociality, LINC00461 with conscientiousness, MTMR9 and WSCD2 with extraversion. These are all traits that play important roles in the development of the cultural tendencies of the different populations and may contribute to the differentiation of the stereotypically interpreted individualistic West from the collectivistic East (Kiaris, 2021). East Asian societies, exemplified by the Chinese, are more hierarchical and collectivistic than the Western societies, and the individuals in the East are typically characterized by higher conscientiousness in their work than those of the West. On the other side, Western societies are individualistic, conscientiousness is typically considered lower, and excitement seeking is more prominent (Kiaris, 2021).

In genes such as CTNNA2, OXTR, LINC00461, MTMR9, and WSCD2, assessment of the frequency of their different alleles in the different populations shows the existence of a good correlation between their versions that encode for what is considered stereotypical behavior in Easterners and Westerners. For example, the alleles that are associated with increased conscientiousness and reduced tendency for excitement seeking, behaviors that are considered more typical for Easterners, are usually higher in Chinese than in European ancestry populations. Similarly, the variations of the OXTR that have been linked to higher perception of loneliness and lower empathy are more frequent in the Easterners than the Westerners. These are traits that are aligned with the collectivistic Eastern societies at which social structures providing support to individuals have been evolved to protect them against stress and anxiety. There seems to be a complementarity in the specific personality traits that these alleles predispose for.

These differences are not restricted only to the extremes, to the populations of the East and the West, but also apply to other populations as well in different geographical locations. Assessment of the frequencies of these and of other alleles that show such "complementarity" in their behavior has shown a progressive increase (or decrease) in their rates in different populations, in addition to those of European and Chinese ancestry, by a manner that follows robust linearity and generating gradients of allelic frequencies (Kiaris, 2022). Thus, in order to establish cultural traits, genes play a role and depend on each other, to fix what we can perceive as a coherent culture. For example, increased conscientiousness combined with also increased excitement seeking may be rather incompatible in the long run, since the former would eventually compromise the latter. Individuals that prioritize seeking excitement would eventually not perform with the same conscientiousness the tasks they are assigned, compared to individuals that do not.

The complexity of the genes' interactions, at the cultural level, increases even more, when additional factors are considered regarding the interactions of the individuals. Social network studies depict the interactions of the individuals in a society, illustrated with dots connected by lines constituting nets. Social network studies in which the genetic identity of the individuals was known showed that the gene DRD2, which regulates the activity of the neurohormone dopamine, is associated with homophily because different alleles of this gene cluster together individuals in social networks (Fowler, et al., 2011). In other words, individuals with the same DRD2 version tend to maintain more social connections with other individuals with the same

DRD2 version. The opposite happens with the different alleles of CYP2A6, a gene associated with detoxification of substances such as nicotine: Whereas DRD2 exhibits homophily, CYP2A6 shows heterophily with no, or rather few only, individuals with the same CYP2A6 version being socially connected (Fowler et al., 2011). The type of CYP2A6 or DRD2 genes we carry impacts the friends we select to have, with the first one selecting against and the second favoring having friends with the same type of CYP2A6 or DRD2 gene.

In different populations, exemplified again by the Easterners and the Westerners, both CYP2A6 and DRD2 alleles have different abundances in European- and Chinese-ancestry populations, which would likely influence the tendency of the corresponding individuals to establish social connections (Kiaris, 2022). For DRD2 for example, that in Europeans the one allele exhibits strikingly different frequency than the other (about 85% vs 15%), connections are predicted to be more abundant. In Chinese ancestry, where the two alleles have frequencies of about 58% and 42%, social connections are predicted to be fewer. In the context of culture and societies, which developed over long historical periods, one would predict that when the connections among individuals are abundant, like in the Westerners, they would supersede the coherence of the hierarchical structures of the corresponding societies, promoting individualism. A dense social network in that case would compromise the hierarchies of the society. In the East, the integrity of the hierarchical social structures is less likely to be compromised, however, because of the fewer social connections among the individuals, which in turn favors stability and collectivism through the establishment of higher rigidity in the societies.

Of course, the contribution of genes such the DRD2 or CYP2A6 gene would only be minimal in determining cultural trends, and these are only two genes that may have an impact, in addition to other genes that may play roles, plus the geography, or the historical events that have shaped how the different populations developed and established their cultural characteristics. But even if the genes play a tiny role in the behavioral characteristics and the establishment of cultural trajectories, it seems likely that certain polymorphisms, by coexisting in different groups of people, depend on each other.

The common good was re-invented. In that case it is the common good of different alleles, that coexist more frequently in different populations. The beneficial effect of this function is not seen in the individuals, though, but rather at the level of the whole population; it presents as a tendency for personality traits that, in turn, when seen together, result in the evolution of certain cultural characteristics. This dependency that is responsible for the beneficial effect is based on the assumption that the corresponding populations will be able to establish cultures that are more coherent, and thus stable, than others.

Apparently, these are all the result of founder effects, from the very early times of human history, when different groups of people migrated, out of Africa originally, and colonized through different steps and genetic bottlenecks, different regions. It is plausible that combinations of genes that provided complementarity in behavioral traits, such as low conscientiousness and high excitement seeking, may act synergistically, benefitting the corresponding population that carries them at such frequencies.

Kin selection and altruism

Very early in time, biologists have realized, the over-arching theme of classical evolutionary theory—"survival of the fittest"—suffers. Frequently, various examples from the study of natural populations have shown that individuals dedicate their own resources for the benefit of others. This actually was happening more intensively when the others were closer genetic relatives. The parental investment towards the offspring is such an example that is seen commonly in nature. In insects, such as bees and termites, which establish communities in which only few retain the privilege of reproducing, is another example. Altruism occurs and has high variation.

The evolutionary biologist WD Hamilton (1964) established the genetic basis of altruism when he suggested that it becomes selected as a genetic trait when the benefits to the recipient exceed the costs to the actor, when the former are expressed as a function of genetic relatedness. From this perspective, it is worthier to sacrifice oneself for a close genetic relative than a distal one, so that the chances to carry your genes will be higher. This formed the basis of the sociobiology argument of Wilson (1975), although later he recognized that genetic relatedness may not be as important as originally thought in driving sociality, especially when it is applied to humans (Wilson & Hölldobler, 2004). Culture in that case, appeared; and culture surpassed what was instructed by genetics. Sacrifice in that case should be viewed not only from the point of one's death that benefits others but also from the wider perspective of allocating resources, and therefore depriving oneself of these resources, to benefit others. Resource deprivation in that case is sacrifice that, although it does not cause imminent death, reduces the chances of survival in the long run.

Two genetically unrelated individuals share the same genes, albeit their exact versions may be slightly different. Both individuals, if they are human, have about 20,000 genes in their chromosomes; they are of about the same size and are in exactly the same chromosomal locations. If you sequence the genome of one, then you will know, with relatively high confidence, the sequence of every other as well. It is their specific alleles, however, that will be different because they will have different polymorphisms and those are the ones that generate all the genetic diversity we recognize among different populations.

One person's alleles will be 50% identical to those of their siblings and 12.5% identical with the alleles of their first cousins. Therefore, kin selection should operate at the level of the alleles, which should favor the versions of themselves that are identical to their own. An individual, from that perspective, does not favor everyone that has hair—because all have hair since they all possess the gene for hairiness—but specifically the ones that have black hair, if they had black hair as well, which makes it likelier to be genetically related with them.

To hold true, this argument of kin selection as a driver for the evolution of social behavior assumes that genetic relatedness is comprehended not only at the level of the genes, as widely perceived, but rather at the level of their versions, the alleles. Each allele should be aware somehow that an identical version of it exists in another individual. This raises another important point as to how one recognizes its relatives so they can make the decision and the commitment, to benefit them at their own cost. One plausible answer is by memory, through learning. You know that these are your relatives because you grew up with them. If you are a tiger, they were also in the nest with you and as soon as they fall out of the nest they may not be recognized as relatives anymore. In other cases, in which relatedness cannot be viewed as a learnt experience, more tangible biological constituents should be present to drive this recognition that operates beyond the learnt experience. Numerous examples support either possibility and extend from microbes to mammals (Gardner & West, 2007; Green et al., 2005; Scott, 2024; Strassmann et al., 2011). Nonetheless, beyond kin selection, altruistic elements in gene function that cannot be explained by relatedness alone are present at the level of the genes. They rely on the hierarchies that are established and by which the benefits from the collective gene action are dispersed disproportionately in all the gene members of the same genome, despite the fact that they may not contribute effectively towards its perseverance.

For humans, at the level of individuals and populations, we can also infer that the learning component becomes more prominent and extends beyond the genetic relatives. The cultures of care and empathy that we have developed instruct us to aid individuals of the same group, and expand over time the size of the group we are members of. Furthermore, we care not only at the level of the resources we allocate to others but also for their overall well-being, beyond what biology can instruct. This is something that is being done systematically. Since the earlier stages of our human culture, it is tightly linked to the characteristics of our species, and it represents the point at which we start appreciating that the distinction between nature and nurture may be deceptive.

3. Caring ancestors

Craniosynostosis is a life-threatening birth defect that is caused by the bones of the skull joining together earlier than they should, creating problems in the developing brain. It is treated with surgery allowing children to live a normal life and have a physiological lifespan. A skull was discovered in Spain (designated as Cranium 14 from the Sima de los Huesos in Atapuerca) that suggests the individual suffered from craniosynostosis (Gracia et al., 2009). The skull is estimated to be that of a child above 5–8 years old but below 18 years old; it belonged to an archaic human species (not *H. sapiens*), maybe *Homo heidelbergensis*. Without appropriate care it is rather unlikely for an individual with craniosynostosis to reach that age, suggesting that in these very early tribal *Homo* communities, children with pathologies received treatment and care similar to those without pathologies.

This is not the only evidence of care in the communities of protohumans. Neanderthals were the dominant protohuman living in Europe and Asia between 300,000 and 30,000 years ago. Analysis of two Neanderthal individuals that were found in South-West France and lived 50,000–60,000, and 43,000–45,000 years ago revealed evidence that they had received extended health care (Tilley and Cameron, 2014; Tilley, 2012; Spikins et al., 2018). The first one, designated LCS1, was a male 25–40 years old; he had extensive periodontal disease, severe degeneration of the lower thoracic vertebrae and arthritis. As such he would have been unable to take care of himself and required the assistance of others for tasks such as securing food or for mobility. LF1, another male, 40–55 years old at death, besides arthritis and periodontal disease had suffered from a fracture of his right femur, transiently limiting his mobility, and periostitis on proximal and distal surfaces of upper and lower limb bones, which is consistent with hypertrophic pulmonary osteoarthropathy. The latter is indicative of heart disease (Assis et al., 2011). Obviously, these individuals would not have been able to survive with these chronic conditions unless they were taken care of by their community.

Care among Neanderthals involved not only assistance in securing food, and mobility, but also access to medicines. Genomic analysis of the dental pulp of a Neanderthal excavated in El Sidrón cave in Spain provided interesting information regarding their diet and also the pathologies they suffered from (Hardy et al., 2012; Weyrich et al., 2017). Among the findings were DNA sequences corresponding to the poplar plant that produces salicylic acid, the active ingredient of the painkiller and anti-inflammatory drug aspirin, with sequences of the natural antibiotic producing *Penicillium*, and sequences of the intracellular eukaryotic pathogen *Enterocytozoon bieneusi* that in people causes acute diarrhea (Weyrich et al., 2017). Putting this evidence together suggests that this particular Neanderthal individual suffered from an infection that caused diarrhea and he treated it with aspirin and antibiotic. Since discovering the medicinal properties of poplar and *Penicillium* by this same individual is rather unlikely, it is apparent that a collective tradition or culture had transferred this empirical information from one generation to another and so was available to be utilized by this Neanderthal in Spain. In that case the benefit did not come directly from the community taking care of the individual by allocating resources, but from the collective information that allowed him to utilize the knowledge that had been generated and transmitted through the generations. Thus, the seed of a healthcare system of knowledge had already been established in the form of information that was accessible to the individuals on a need basis.

Besides a form of embryonic healthcare, the communities of the Neanderthals attended additional requirements that extend beyond the basic biological needs for their survival. Analyses of the sites where the 50,000-year-old Neanderthal remains were found indicated that they practiced burial of the dead (Rendu et al., 2014). How extensive this practice was remains unclear, but it shows that the care of their group members extended to the afterlife as well. Even more impressive were the findings at a significant Paleolithic site in Shanidar Cave in Iraqi Kurdistan. At this site, men, women and children were buried and besides evidence of care towards the group members who had suffered from injuries or other pathologies, pollen was revealed around a dead body—that is indicative of flowers—and it is also suggestive for the implementation of a ritual burial practice (Pomeroy et al., 2020). Furthermore, comparative analyses of the different remains suggest that the Neanderthals were returning to the same location to bury their dead, revealing the significance of certain localities for such practices. Such practices require the allocation of considerable effort and resources. It becomes apparent that already, in archaic humans, the needs were acquiring a wider perspective from what had happened so far for the animal kingdom and involved spiritual practices. In other primates, such practices and rituals do not exist. Nonetheless, seeds for the perception that a dead body may represent something beyond mere inorganic matter may have existed before the protohumans. For example, chimpanzee mothers may carry the bodies of their dead infants for up to 3.5 days, even though behavioral evidence shows that they recognize that the death has occurred (Lonsdorf et al., 2024)

Evidence for care and empathy are more abundant among prehistoric *H. sapiens*, not only because they were becoming more common over time but also because they are more recent in history. Several examples are available from excavation and analyses of the remains of prehistoric humans. In one of them in Northern Vietnam, the remains of a greater than 3,500-year-old male was found who had suffered from quadriplegia acquired as a young child (Tilley and Oxenham, 2011; Oxenham et al., 2007). The individual was apparently paralyzed and unable to attend his own needs. This represents a very early case of survival despite severe disability that was possible only due to dedicated care. A similar case of care in prehistoric humans involves the 7,500-year-old remains of a 15-year-old male found in Florida that suffered from spina bifida aperta, or open spina bifida (Dickel and Doran, 1989). This is also associated with severe neurological problems and is indicative of a condition that requires intensive care by others for the individual to survive and reach the age of about 15.

As we move forward in time, and even before societies became structured by the complex ways that are consistent with what we perceive as civilizations, abundant evidence indicates frequent deformities in skeletons from burial sites, such as those in Spain of 800–400 BCE (Rohrlach et al., 2014). These cases revealed evidence of chromosomal abnormalities, suggestive of even more severe conditions in which disability is not only physical but also mental. For example, evidence of Down syndrome due to trisomy in chromosome 21 and Edwards syndrome due to trisomy 18 was found. Despite their physical disadvantage due to chromosomal alterations, the individuals were still considered to be members of their communities, and because of that, they deserved care during their lives and were treated accordingly, after death as well. Not only they were not abandoned because they could not contribute, but they were taken care of, during their lives, and were buried together with the other group members when they died.

These are just some of the several examples that show that very early in human history, even before our species' evolution, in protohumans such as in the Neanderthals and probably *Homo heidelbergensis*, and soon thereafter in prehistoric humans, our ancestors cared for the members of their community and provided aid and support to those in need. This care spanned food provisions and carrying them when they became immobile, and extended to the development of traditions that passed from one generation to the next regarding the beneficial properties of certain plants and extracts, and even involved care about their afterlife and their proper burial. The line between the biological needs and the needs instructed by the embryonic culture that started developing, becomes vague. From a biological perspective these individuals that suffered from pathologies, injuries, or chromosomal aberrations could not contribute to the group now or in the future with their own effort and resources, nor could they contribute to its genetic pool with their reproductive potential that was compromised. Nonetheless they were taken care of, not simply because of their past effort before the disability, but just because they were group members, and they were entitled to this care.

Already, a culture of care had emerged according to which resources, in the form of provisions and effort, were allocated to members of the community in the absence of a trivial biological benefit to the care givers. Such culture started, or rather likely co-existed, with the protohumans apparently independently in different geographical locations, became more prominent over the generations and represents a major and systematic characteristic of the upcoming civilizations that appeared on Earth. It becomes apparent that the group's efforts start orienting towards the defeat of the biological limitations that certain group members may have had since birth or acquired after their birth.

4. Civilizations: Social structures assisting everyone, almost...

Complexity emerges

From about 200,000 years ago when our species evolved, and for the longest part of our history, humans were hunters and gatherers, living with what was available from them from nature without making special provisions to take care of their needs well in advance. They did have tools to make their attempts effortless and more efficient, and they were also developing small social structures in the form of tribes so that by cooperation they could achieve more than they would have achieved by working independently. Then, around 10,000–12,000 years ago, humans made an invention that had a profound impact in their future: They started developing strategies and techniques to domesticate and culture plants, in other words they invented farming and agriculture. At around the same time, or maybe even earlier than that, they discovered how to domesticate animals as well, living as nomads and establishing pastoral communities. This was a great advancement—but not of the scale and impact of plant domestication. Agriculture had very prominent implications in their societies because it had a more consistent effect in their size and their complexity, due to the abundance of resources, the surplus that was generated that, in turn, mandated the development of bureaucracies for its management, and the coordinated efforts needed for its protection and growth.

When and where agriculture first started remains unclear. Some believe that this may have happened almost simultaneously, in different places of the world such as Mesopotamia and Egypt, China, or the Americas. The domesticated plants, after all, differed in the different places, with wheat being the prominent plant along the Tigris and Euphrates rivers in Mesopotamia, rice in the Yangtze River Valley and Yellow River in the East, and maize in Mesoamerican planes and lowlands. It also remains possible that it had started in a single location, likely in the Fertile Crescent of Mesopotamia, in the borders of modern-day Iran and Iraq where the earliest evidence indicates the practice of agriculture, and then spread around the world where it would have been applied to local plants that have already been adapted in the corresponding regions. While the actual plant species are different, the strategies remain similar.

Irrespective of the specific details, though, agriculture caused great changes in how people lived and organized themselves. The profundity of this can best be grasped when we consider that the period before that, starting from the point at which our species originally evolved, accounts for more than 95% of human history, while the period after the advent of agriculture encompasses just about 5%, or less, of human history. The changes were dramatic and can be appreciated by comparing the progress attained before and after agriculture, in the first 190,000 versus the latest 10,000 years of human history. From the mobile settlements of tens or hundreds of foragers, people started forming cities of thousands, which soon became larger and larger, reaching the size of more than 50,000 occupants, such as in Memphis in Egypt around 2,000 BCE, or the Sumerian cities of Lagash and Ur, around the same period. Positive feedback dynamics were set in motion since farming made possible the rapid increase in the population, and at the same time also made human populations become more immobile as the people now had to stay around their crops to protect them, feed them and harvest them. People could not just move around anymore in the face of adversities, as they had done in the past as foragers, but they had to stay close to their wheat, their rice, or their maize cultures, which in turn were increasing even more their population and enhancing their interdependence.

Agriculture also created a surplus, that enabled population growth and induced inequalities in resource distribution. It also demanded the development of bureaucracies that were tasked with resource management, and this established hierarchies. Furthermore, people now had to protect their crops and the fertile land from other people who may have wanted them; thus they had to develop armies. Skills also started becoming more specialized, such as the skill of knowing when and how to harvest (and what), and also how to fight. This automatically added another layer of complexity. Different skills were demanded for the function of the society—skills which had to be taught, not to everyone but to the ones destined to practice these skills, and their practice was not always applied evenly throughout the year. Soldiers, for example, had to be trained and to maintain readiness during peacetime, while the ones that knew when and where to harvest and make more strategic decisions for the group were active at specific periods as well. Furthermore, more food and wealth were produced than was needed, which put forward issues of distribution. Thus, various provisions had to be made to coordinate these activities and the distribution of the surplus. This eventually caused the development of much more complex administrative systems and the bureaucracies associated with them; the establishment of hierarchies that besides their different roles and specific functions also had different access to resources; and ultimately, in the emergence of states as we recognize them today and record their legacies through the historical and archeological evidence. These states now started enforcing group membership based on birth, while earlier it was more on a voluntary base. As a forager, you and your companions could easily leave the group if you wanted to, living as a forager elsewhere if you were willing to take the risk of abandoning your current group. In the states of the agricultural societies, though, your birth automatically reflected your roles, your presence in the group and your social trajectory within it. The notion of "good citizenship" started emerging within this context.

According to some recent views this complexity in social organization may have occurred independently of the widespread practice of agriculture, with some prehistoric societies already becoming complex without being agricultural. In other instances, agriculture was practiced without necessarily resulting, and being dependent on, complex hierarchies and administrative bureaucracies, at least at its first instances (Graeber and Wengrow, 2021). It seems though that this recent period, of the latest 5% or so of our time line as species, was rather transformative for mankind. More coherent cultures and eventually civilizations started emerging. Irrespective of whether complexity and hierarchy are inherently associated by causality or not, from the time point of a few millennia ago we start seeing the formation of cultures that develop themselves as civilizations. These civilizations emerged at different locations in the world, matured, and eventually collapsed and died. Why and how they did so is now speculative. It could be due to wars and conflicts with others that absorbed them, or they may have disintegrated over the course of their histories. In no case though, so far, has a given civilization transformed into another one by the same people; such transformation was always associated with new people bearing new ideas, mind sets, and collective imaginary that fueled the disappearance of the former and its substitution by the latter. This is what offers a coherence in what we now perceive as past civilizations; and despite their cross-fertilization, we view them as single and unique entities that are defined by and correspond to given peoples of the past. These civilizations, if we compare one to another, were characterized by several differences in their fundamental structures and functions, and the ways that people identified themselves within them and were integrated in the corresponding societies. Furthermore, and foremost, people recognized their continuity in the cultural heritage they had inherited. They were starting their lives with a specific background that relates to their past, and this represents an indispensable component of their identity. Nevertheless, despite their specific differences, these civilizations also had similarities that to be appreciated: we should consider that they had developed largely independently from each other and that their identification may point to common aspects of human nature that can be quite illuminative regarding our collective future.

One of these civilizations, the one that we currently live in, is the Western civilization, which we will discuss more deeply in the following chapters. Another civilization that is contemporary and started at about the same or maybe slightly earlier than the Western civilization, more than 3,000 years ago, is the Chinese civilization. Until a few hundred years ago the Chinese civilization was unfolding independently of the Western civilization and it remains alive and vibrant until today. In modern times, however, the two are in the process of merging to a large extent, and according to the most prominent views it appears that the Western civilization is absorbing the Eastern. Whether this trend will hold in the future or, due to geopolitical factors, the Eastern civilization will become more dominant, only time will show (Frankopan, 2019).

At least according to some, and probably the most traditional, views, human inequality traces its roots to the social changes that followed the spread of agriculture and is a direct consequence of the surplus that was generated and was distributed unevenly between the people. The development of states was also intrinsic to the widespread practice of agriculture. Cultivating crops causes and depends on the significance of localities and the borders that had to be defined, safeguarded, and protected. Thus, administrative structures had to be developed to regulate such borders at the level of the individuals and in relation to the authorities that own them, and in relation to the whole population versus others, the outsiders. These administrative structures, which evolved into hierarchies, were also responsible for allocating resources and distributing the wealth that was produced.

They also regulated spiritualistic concerns related to religion and the divine in a process that also re-enforced and legitimized their authority.

Culture now starts becoming intermingled with the notion of the states, from the perspective that the former fuels the comprehension of an identity that is necessary for the operation of the latter. Not that states require cultural homogeneity in order to operate and to be successful—they only need comprehension of the hierarchies and the acknowledgement that to a variable degree, this is related to certain benefits of group membership. Several paradigms indicate that cultural homogeneity is not a prerequisite of a successful state, and they point to the fact that actually, the most successful ones were culturally heterogeneous: for example, the Roman Empire, a very successful and huge state that was based on the cultural heterogeneity and tolerance of its citizens. Another, earlier example, comes from ancient Egypt, during the Middle Kingdom around 16th century BC when Egypt was a multicultural empire, with the Nubians in the south being a distinct cultural group (Smith, and Buzon, 2018).

One has to consider two elements to appreciate why multicultural empires could become successful and persist over time, notwithstanding the fact that forces with different interests and priorities may co-exist within them. First, that effective governance is not always attained voluntarily, by will that is based on cultural underpinnings, but also—and indeed in most instances—by force. Second, that culture and civilization are not equivalent and interchangeable notions. Different cultures, despite the fact that one is always more prominent, may coexist in different civilizations. More importantly, not all members of a state should carry the same culture. Heterogeneity in cultures and their tolerance can very well be a characteristic of a civilization, as long as some fundamental ideals and social imaginaries are followed. Those should primarily focus on the comprehension of the specific hierarchies and the establishment of common imaginaries that will be able to engage and motivate people towards a mutually recognized goal. In other cases, different cultures that co-exist under common collective imaginaries progressively unify by sharing more and more of their distinctive properties. The Western civilization represents such an example in which peoples of originally different cultures, around the world, engage progressively into a new more homogenous culture that relies on a capitalistic foundation and in which the Western ideals acquire increased prominence.

Some common themes can be recognized in different states, irrespective of whether they existed in the different civilizations. Societies started exhibiting increased division of labor between their members, with some individuals, usually the majority within a population, being charged with tasks relevant to food production, others being assigned to the state's protection (or expansion), others participating in the administrative bureaucracy, others serving as priests or healers, and so on. In all cases and independently of whether we focus on the Sumerians, Babylonians, Egyptians, Chinese, or Incas, different tasks were assigned to different individuals in a manner such that hereditary privileges or burdens played a major role. Over time, though, and albeit with tremendous variation in different cultures and civilizations, personal qualities and characteristics had their impact as well. This was more apparent in tasks that were considered pre-eminent and vital for the state's operation, such as the population of the military class. In some cases this reached extreme positions, such as in Sparta where the whole purpose of the society was to produce and train the Spartan soldier. For those who were not deemed capable of becoming one, there was no place in the society. In Sparta, boys that would not make good warriors were sacrificed. As far as tasks that were considered to require enhanced intellectual capacities and training, in ancient China for example, national examination systems were established as early as 2,000 years ago that aimed to populate the administrative elite, again with individuals having certain capacities and promise, who were recruited for appropriate training. The early inception and implementation of the national examination in China has been recognized as a major contributor to the increased homogeneity compared to the cultures of other civilizations because it trained all its participants, at least 1,000 more than those who succeeded in the exam, in a distinct and defined culture (Cressey, 1929).

An increasing specialization started developing for different tasks and the individuals that were charged to perform them. This constitutes a tremendous difference from the past, the pre-agricultural period when each individual could, and actually should, be able to do practically everything himself. He should gather the food, take care of the sick, actively engage in the protection of the group, and also participate in the collective decisions that had to be made, giving rise to more egalitarian structures than those we are used to today.

The post-agricultural societies were not characterized by high social mobility, and frequently they did not even aspire to do so. Where and to whom one was born determined how people would live their lives. Nevertheless, provisions were made for the appropriation of certain people with specific characteristics and qualities and their assignment to certain tasks. This, starting from the earliest points of social organization, was eventually associated with the introduction, establishment and strengthening of inequalities because it was deemed that different groups should be rewarded differentially in each society; or they had the means, because of their status, to reward themselves accordingly and in parallel with that to sustain a tradition that rationalized these choices. Usually, it was a combination of factors, such as hereditary privileges, randomness, or merit, that guided, established, and re-enforced inequality in the different societies. An elite always enjoyed disproportionately more of the privileges, in a manner that was justifiable, or at least acceptable, settled, and unquestioned by the society. The distribution was guided by heavenly mandates or, more recently, by the established societal norms that guide the economy and corroborate what is now perceived as a well-functioning society. The difference is that today, the sustained and indeed increasing inequalities in wealth distribution unfold against the backdrop of aspirations to equality: a widespread belief that people have the opportunity to become members of the elite based on merit and effort, a widely accepted ideal of the Western tradition. The reassurance of this promise and the strengthening of this ideal that culminates in the American Dream resurfaces vigorously before the elections and contributes to the maintenance of social harmony despite the dramatic increases in inequalities in the last decades.

Choice and chance in group membership

What made people across history and throughout the world in all different instances choose and live in societies characterized by high inequalities, when statistically, the chances of being underprivileged is higher than that of belonging to the elites? This is a difficult question to answer, and its actual meaning may even be quite obscure since the term "choose", in its traditional sense, is not applicable. No person really is ever asked, after he or she is born or becomes adult, whether he or she would like to live as a forager, participating in the collective decisions and enjoying, more or less, the same privileges with the others of the group, or to live in a post-agricultural society which is characterized by increasing inequality. Probably, an exception can be found in Ancient Athens, and after the Enlightenment in the West, when such inquisitions, at least at a theoretical level, were active and ongoing, suggesting that people do believe that they could be organized differently. They did not change the society they lived in but they could easily project themselves as living in societies that functioned differently. Nevertheless, the fact that eventually human societies, in their vast majority, converted in the direction of unequal agricultural societies implies the acquisition of a collective benefit. One would assume that if such a trend were not functional then the corresponding societies would have become unstable and would have perished, but they did not. Thus, there must be a certain advantage that makes societies in which the few enjoy disproportionately more than the many become increasingly prominent in history.

The use of the term "choose" should also be seen from a different angle. It does not possess the meaning of a conscious decision that has to be made when someone is faced with two plausible outcomes, but it is similar to the "choice" made during biological evolution. Birds did not choose to have wings so they can fly, but wings were chosen over other adaptations because they provided their bearers with certain advantages. In that case the choice is not made by the individual but by the evolutionary process dominated by trial and error. In a similar manner, when it comes to societies and the choice of how they are organized, it is their effectiveness that made the choice; they might persist, or disappear, because of the maximization of the benefits they provided to some individuals in combination with their minimization, that is the disadvantages, it provided to others.

To understand this benefit, and to recall philosopher John Rawls and his theory of justice (Rawls, 1971), we should probably try and see ourselves in societies that operate behind a "veil of ignorance". In this condition we will have to be naïve regarding our capacities, our characteristics, and our heritage, and we will have to pick the society we would like to live in, ignorant of where exactly within it we will eventually land. We could be among the elite that enjoy the most, or the peasants that enjoy the least. We could also be foragers. All we can choose, however, is whether there will be an elite and how this elite and the peasants will live, and maybe, how many they will relate to each other. Considering also that societies change progressively over time, we should make choices that do not differ drastically from the current or a previous state. In other words, if we are living in the feudalistic Europe of the Middle Ages, we could maintain as a valid choice the proto-capitalistic economies of Northern Italy of the 13th century. If, alternatively, we live in a pre-agricultural society, we could just opt to keep on living as foragers—or we could participate in the early agricultural society. In our choices we should consider only the current state of the society and class as static, while also assessing the prospect of change that may be present.

Granted the oversimplification of this analogy, it may inform about the readiness of the feudalistic societies to make changes, because of the accumulated discontent. It could also inform about the plausible will of foragers to be involved in agriculture. Our decision and what the majority of the people will choose will inform about the society's dynamics and whether it reflects an unconscious choice supported by the people or it is destined to change in due time. In that case, the answer becomes very simple, and it will be based on the principle that it will be a society that is able to attend, at its minimum, to the basic needs we will have, as disadvantaged, if we happened to land in this class. It remains of course to define what this minimum entails, and we will try to do so in subsequent chapters. In parallel it should also be a society that provides more to the ones that had received the minimum in the earlier stages. This introduces the notion of anticipation, that in turn justifies and fuels what is perceived as progress. More recently, studies involving more than 6,000 participants have shown that reasoning based on this "veil of ignorance" results in the greater good (Huang et al., 2019). In view now of the fact that the greater good is highly charged in relation to the historical time, the potential deviation from the greater good reflects the instability of an early society as compared to the one that may replace it.

In these structural, hierarchical and unequal post-agricultural societies, always, each and every individual started out having its essential biological needs taken care of by the society. This reflected the benefit that was appreciated and that contributed to the society's stability and justified the "choice" over other types of organization. Thus, behind this veil of ignorance, collective reasoning was applied that benefited more the ones in the worse condition and were most likely the majority. Naturally, what defines the biological and essential needs is rather vague and arbitrary, and at its minimum must have included at least food and shelter, and protection. If we want to see that from a more biological perspective, note that all these converge into the assurance of our efficient reproduction, our intrinsic need to have offspring that will also be able and reproduce and give our genes the opportunity to persist over time. Pre-agriculturally, uncertainty about food and resources was perceived as the limitation that fueled the social transformation to a state in which inequalities were magnified.

The requirement to make this transformation was morally justified and the demand for this requirement to be taken care of, by the hierarchical elite, was considered not only fair but an imperative. It was actually the single most important charge that the governing elite had: to take care of its subjects. In all societies, from Mesoamerican and Mesopotamian civilizations to Ancient Egypt, everybody had a role so long as the society considered that he was doing what he was supposed to do so. In return he or she should not starve, should have a place to live, and should be protected if his household faced dangers from external forces, or even internally, if they faced aggression from other members of the same society. This ultimately translated into securing the ability to have a family and to have offspring that would enjoy the same privileges. Codes and laws were being developed that safeguarded the society's operation, and although for most of our history this power of the ruling elite was unlimited, it bore a component of subjective morality that justified its establishment, its function, and its maintenance in this position. This is typically reflected to the Mandate of Heaven in Chinese history, according to which every ruler governs because he is doing his job right, and thus Heaven favors him. If a usurper appeared, as happened when dynastic authority changed, it did so because the previous ruler did not fulfil his obligations and therefore he lost the Mandate. A wise and failsafe resolution indeed in which neither outcome may be seen as unjust.

In these hierarchical and unequal societies, albeit highly disproportionately for their work effort by comparison with today's standards, everyone should have been able to live until maturity, reproduce, and have offspring. These were the dynamics that established the transformation of the late pre-agricultural societies to the more complex and unequal early agricultural societies. Of course, another factor that played a major role is that of dominance and violence, according to which individuals did not "choose" but were rather "forced" to live under these newly incepted conditions and were discouraged, by the application of violence, from living differently. Yet, since there was never an active choice to be made, the question is reduced: it is not so much why people selected this type of society as why these societies persisted over time.

An interesting twist emerges here that appears to contrast with what happens to other species from a biological perspective. Agricultural transformation led to a significant increase in population. This is consistent with the biological priorities whereby the efficiency of the new strategies is gauged by reproductive efficacy. Nonetheless, when status is considered within the agricultural societies, the number of children does not directly reflect the individuals' social and financial status. Furthermore, the privileges enjoyed by the elite could not be attributed to their biological abilities. In other species where differences in social status exist, they are associated with reproductive success (D'Amato, 1988; Ellis, 1995). Yet people want to have higher status despite this not being the pre-requisite for having more children. Furthermore, status is not indicative of either intellectual or physical abilities. It emerges as a state that is desirable, but it does not translate into long-term biological benefits. It only implies better well-being that does not translate into an increase in reproductive efficiency.

An argument against this point could be that the males of higher social classes and especially those that belonged to the aristocracy could frequently have more offspring, particularly if we consider those outside of marriage or of the established spousal relationships. This, however, was not happening at such a level and extent that it could be justified as being biologically impactful, and there are also several examples in which lower class citizens were having several children just because they were not obliged to conform to the stricter norms applied to the aristocracy. In all cases, though, we can see that an individual, at least biologically, could be successful in terms of reproduction, just by belonging to an organized society, and irrespective of whether he or she ranked higher or lower within these societies. They could live at least to maturity and reproduce, and know that their children likely would also be able to do so. Everyone was, one way or another, integrated in the society. People's reproductive efficiency had been dissociated from their performance within the society and their social roles. The privileges of status do not translate to reproductive success that was warranted for everyone. Importantly, this was also the case for their children, who would find their role in the society just as their parents had done. The desire for and pursuit of higher status was present, but it fulfilled other needs than reproductive efficiency. This dissociation between the function of people within the society and of its organization in different classes, and their efficacy in reproduction, reflects a point at which the impact of culture becomes of particular significance in people's lives.

All the changes that were linked to agriculture stimulated people's interdependence because for these new societies to function efficiently, specialization was needed, and that renders individuals inadequate to meet all their own needs directly. They constantly had to rely on others. In turn, this interdependence stimulated the development of norms that contributed to the coherence of the society and the fixation of the perception of belonging to a group, outside of which the costs outranked the benefits. Precisely this also contributed to the gradual change of what constituted the needs of people and the classification of categories of needs, many of which are not biological but are instructed by the society at a given historical time. Continuously, group membership warranted that these needs be taken care of. Besides the arbitration in defining what the needs are, the definition of group also varied considerably across the history of humanity. The unit could be the band, the tribe and eventually a city state, the kingdom, the empire, or the nation in the most recent years. In other cases, it was, and still is in some instances, a religious group; and all those dynamics were becoming intertangled across history. Membership in the group, in most cases, was not voluntary, and nor was the definition of what constituted an essential need.

Needs: Essential, historical, and desired

Throughout history, in order to live what they perceived to be a "good life," people developed a series of recognized needs that required satisfaction. Such needs changed over time, and their attendance guided how people were organized in the societies they formed, by setting their priorities and aspirations. These needs also defined their compromises and determined their relationship with their leaders and their political institutions.

The needs of individuals can be divided into categories depending on their origin and the historical period. The first category includes the needs that are considered essential and are the ones that directly determine the degree of biological success. Food, shelter, and protection represent the foundation for this first category of needs that converge into people's reproductive success. Taking care of those needs was the primal concern of prehistoric people and so it continues in the later phases of history as well. These are the essential needs within a given society, and when societies became increasingly hierarchical, it was the main task of the ruling class to satisfy them. No matter how hard peasants worked in the fields, whether they cultivated their own property, the emperor's fields, or another landowner's land, it was warranted for them and their families that shelter would be provided, food would be available to sustain them, and in case of attacks by hostile forces, it was an obligation of the rulers to develop military systems to defend them. This did not apply only to external forces but also to disputes among individuals of the same group, in which case the authority was also charged with applying the rule of law, no matter how this rule was perceived. One needed to know that by abiding by this law he safeguarded the state's protection and his share in privileges. All these, taken together, had to make sure that the individuals would reach reproductive age, would produce offspring, and that these offspring would also be able to reproduce. Biologically they would all be successful, and their degree of this biological success would be largely irrelevant to their social status. Of course, healthcare and access to resources, both of which contribute to reproductive success, were disproportionately distributed according to one's status, but the overall goal was to consistently render such resources accessible to more people. The responsibility to make sure such needs would be satisfied fell on the ruling elite that, independently of how the wealth was managed, at least had to make sure that these needs would be covered. The society, personified by its rulers, should make sure that everybody would be able to reach reproductive age and produce offspring. Otherwise, societies will become unstable, and disintegrate, and their function will become compromised. Population growth was always interpreted as success in a society because it followed abundance of food and safety. From that perspective the agricultural societies were extremely efficient because they immediately resulted in population growth (Bocquet-Appel, 2011; Gignoux et al., 2011).

Of course, the increased complexity and growth do not mean the operation of idyllic forms of social contracts wherein the rulers and the subjects reach an agreement on how governance should be done, nor that the implied social contract could be modified voluntarily to satisfy the changing needs of the people. On the contrary. Obedience was demanded that enforced the *status quo*, and only in few historical instances did the elite, usually forced by the ongoing dynamics of the society, voluntarily abandon any privileges for the sake of the lower classes. Typically, the laws were harsher as we move towards the past and punishments were more severe. The Qin Dynasty in the late 3rd century BC, that unified ancient China and set the basis for an empire that lasted for 2,000 years, was notorious for its harsh laws and severe penalties under the doctrines of legalism. Similar to this was the legal code of Draco, in ancient Athens in the 6th century BC, that also established harsh laws, from which the term "Draconian" was derived. In both cases, though, and in subsequent examples in which a central authority empowered itself, access to the essential needs was inherently associated with the sustainability and perseverance of any given social system. Cultural developments and the social norms associated with them assisted in the development of a system of reciprocity whereby all contributed—albeit unevenly—to the society and anticipated receiving back the satisfaction of their essential needs. If such needs stopped being satisfied, drastic changes had to be applied, relieving the tensions. Essentially two choices were possible: either to adjust the existing framework and seek wider acceptance from the society and its people, or to disintegrate. In China, the Qin dynasty was soon followed by the Han Dynasty, a peak point of Chinese civilization. In ancient Athens, a series of events was set in motion that culminated in the establishment of Athenian Democracy less than two centuries later. Analogous were the dynamics in England in the 13th century that resulted in the adoption of Magna Carta and the voluntary abolishment of certain powers of the monarchy in favor of the people. Such matters, in today's Western liberal societies, are absorbed by the political systems we have developed. The same dynamics are applied, according to which the ruling classes are legitimized by the tolerance of the whole society, that evaluates them by considering whether such needs are taken care of and satisfied. Following periods during which the authority of the ruling classes possessed a theocratic foundation, people accepted their rulers just because their main needs, the essential ones, were met. People expected these needs to be taken care of, and the rulers had the obligation to attend these needs. Violence and force were always an inherent component of the authority to establish and maintain itself; nonetheless, force alone would be insufficient for them to persist unless a benefit to the people could be provided. Historical context of course plays an important role in deciding what constituted an "acceptable" use of force and what magnitude of needs should be satisfied.

The second category of needs involves needs that were not associated with the reproductive efficiency of the individual people, but which, at some given time, were also considered essential and just as important as the essential or biological needs. These needs that do not translate directly into reproductive performance allow us, the observers, to monitor and record historical progress and how the collective mindset of people changed over time. Nonetheless, at any given historical time point these needs were always perceived by the corresponding people as equally essential to their biological needs. Their exact nature was determined largely by their history and traditions which were establishing a foundation upon which, each generation of people started its life.

Those needs not associated with reproductive efficiency but which were considered equally important, the historical needs, were the ones that gave the specific character to the corresponding societies. These needs were constantly changing over time, but were always perceived as equally important with the biological needs and their classification as such is only done retrospectively. At any given time in history, they were considered to be fundamental. They were widely accepted by the people and despite the fact that they differed dramatically in different cultures and time periods, they were always considered essential and unquestionable for their significance and role in the society.

A peasant during the Han dynasty in the first century had different expectations of the dynastic rulers, and enjoyed a different quality of life, from the peasant in the Ming period fifteen hundred years later, but more or less, one may assume that the peasants in both cases felt that they were receiving similar degrees of attendance by their ruler or the political system, and both were living in a society that at its peak periods was taking good care of them. The same applies to the ruling class of these periods and their obligations vis-à-vis their subjects. Analogous responsibilities and ambitions guided their function and determined their success and their acceptance by the people, despite the differences in their authority and the particular historical context.

In ancient Greece, a citizen of Athens had drastically different expectations from the state and the expected functions within it, than a citizen of Sparta in the 5th century BCE that was only 150 km away. Yet in both cases the citizens felt the same degree of pride and satisfaction in being members of the corresponding society. It is noteworthy these two opposite forms of social organization, the military Spartan versus the democratic Athenian, were followed by the same Greek people, who spoke the same language and believed in the same gods and developed together, having diverged within just a few centuries.

Most of the time the satisfaction of these needs was supported directly by the government or administration. However, at other times such needs were satisfied by other mechanisms because of the tolerance of the administration. In that case the administrative tolerance had a penetrance in the society, was expected, and constituted an integral part of it. An example of this is with the religious practice. At certain times religion operated by being tightly intermingled with the governing authority such as in the Middle Ages in Europe; it might be merely tolerated by the authorities such as in the Roman period; or it might operate in parallel with the authority, but satisfying specific social needs, such as today in the West. In all these cases, the special needs satisfied by the religious practice that was conceived as essential by those attending it. Irrespective of whether the government formally imposed religious norms, operated at the discretion of the religious authorities, or freely permitted religious practice, the individual citizens perceived the exercise of their religious rights as fundamental. Of course, in several instances, even today, religious minorities existed and continue to exist. Historically, though, their existence marked periods of transformative historical changes that altered the course of history and the fate of the people, until new dynamics were established and new balances were reached.

For the individual people, historical essential needs had a different meaning at different times and at different contexts. As we progress in time, and travel to different regions, we realize how different such needs can be. Nonetheless, in all cases a relationship is established between the rulers (kings, tribal heads, emperors, elected officials) and the subjects or the citizens, by which the governing elite enjoys disproportionately more than what may they directly contribute to the society, and their acceptance as such relies on the tolerance of the people, because it is deemed that it resolves and addresses the peoples' needs. It was always however, the expectation of the ones that were governed, that the governing elite will at least be able and satisfy those needs.

Prosperity and wealth are commonly a historical need that was fundamental for different societies and reflected the well-being of people. We have a quite good understanding of these needs, from examining the today's societies and by comparing the perception of prosperity for a worker in different time periods. By comparing the work schedule and the compensation of a factory worker after the industrial revolution in a Western country, we will come to an appreciation of how relative such expectations can be. Originally, workers were expected to work for about 16 hours per day, 6 days per week, while today it is less than 8 hours per day on average for 5 days. Work conditions, associated benefits, and of course compensation, were also disproportional throughout the years. In all these cases, however, work was voluntary, and people chose to work in a factory in big cities rather than farms in the countryside. Obviously, they found some advantage to working as industrial workers, despite the differences it entailed in different periods, and this sustained the industrialization of the last centuries. One can naturally argue here that the inhumane conditions, by today's standards, of the factory worker of the early 20th or the 19th century caused the series of the drastic transformations that resulted in the more "humane" conditions of today. Furthermore, it was during these periods in the industrial revolution that new political theories emerged, examining alternative ways by which people should be organized, and raising systematic doubts about the existing forms of socio-economic organization. These were either put into practice or drastically influenced the existing system. Nonetheless, they reflect a state in which the given order of things no longer meets widespread acceptance by the people and some reorganization must incur, as it did. A counter-argument of course can also be made: that similar social turmoil also exists today, even in the most advanced Western societies, and although the challenges emerge from different targeting groups making different specific demands, they will lead to analogous changes as well.

Societal needs are not only biological

The needs for material goods increased over time, simultaneously with the invention of more efficient ways of production. More food and more efficient technologies to address healthcare, shelter, or mobility were becoming available and their products were distributed to more people. These were not, however, the only needs that the people demanded should be met. Since very early in human history, people have appreciated that a good life translates into more than biological success, and this is an inherent component of our cultural evolution during which reproductive performance became disentangled from status. It also entails other components that we collectively define as culture, which in its own right has different facets and dynamics.

Frequently, the peoples' needs that integrated into their well-being, in its widest sense, were viewed in the sphere of the metaphysical. In China, since the very early times of its history, as expressed in Confucian teachings, rituals and sacrifices were an indispensable component of the rulers' obligations and were a necessary element to secure the Mandate of Heaven that legitimized their authority. Yuan Shikai, to legitimize his monarchy and new dynasty in 1915, following fall of the Qing, performed such rituals at the Temple of Qing. In ancient Egypt, and with some variations in the Western world during and soon after the Middle Ages, rulers also legitimized their authority by being recognized as the representatives of God(s) on Earth. At the same time the obedience of their subjects was a necessary component for the well-being of their souls in the afterlife. Today too, such satisfaction of afterlife concerns, combined with morality codes and the perception of a life well lived, is associated largely with how religious practice is integrated in the society. How the rulers accommodate such societal needs changes with time. In the West today, it is not the authority's duty to satisfy such needs, but it is their duty to protect the right of the people to do so by establishing the legal framework that permits it and protects it. More recently, and even today, in some Islamic nations, the intermingled divine and secular authority promises a life well-lived and mandates obedience by warranting those appropriate actions. In these societies suffering and renunciation of earthy pleasures and biological necessities, including life itself, are all justified in exchange for well-being after life ends.

Cultural functions are also greatly supported, and frequently used, by the state for its own goals. Today, almost all European countries have ministries of culture or the equivalent that are charged with the mission of satisfying the cultural needs of the society, recognizing culture's role in people's well-being. At various times, such as during the Roman Empire, the authorities also supported the production of spectacles, a policy that culminated in the famous "bread and circuses" strategy that was used to enhance their dominance over the people and distort their sense of satisfaction. Before that, in ancient Athens, the city-state coordinated competitions and supported the production of several tragic plays to be staged in the festivals of the Greek classical world. In one of them, in Dionysia, Aeschylus won the first prize in 472 BC.

It becomes apparent that since very early on in human history the state and its rulers adopted as an essential and indispensable function the cultural entertainment of the people in a process that frequently acquired the reciprocal meaning of both reflecting the needs and guiding them accordingly. In Plato, elements of such state-sponsored art and of its political engagement exists, according to which the rulers should use the power of artistic creation to cultivate people in order to educate the ruling class, from very early in their childhood. Totalitarian regimes also frequently utilize the power of art in guiding peoples' goals and mindset. Similar arguments also exist today against the mass culture of capitalism, exemplified by the pop culture that was strengthened in the 20th century, that is promoted to support the system's perseverance by guiding and formulating the peoples' desires, preferences, and aspirations (Southerton, 2011).

There is a huge diversity in the ways that various historical needs became essential, in a process that is aligned with the collective imaginary. Besides religion and art, other aspects of peoples' lives become essential as well. Now freedom of speech and the circulation of ideas are also considered principles that are fundamental components of human existence and are considered essential for the people's well-being. In this case, of course, the distinctive line is rather vague because legal protection of such rights may be different from its widespread acceptance by the society. In the U.S., for example, where the circulation of ideas is free and protected by the Constitution, different social groups at different locations, workplaces and social environments, have developed different degrees of toleration for various ideas and behaviors, although all are legally protected. When these lines are written the clarification of what freedom of speech in the University setting and the notion of institutional neutrality associated with it have emerged as major issues of discussion and controversy in American academia. Outside the academic institutions, in various communities certain borders of tolerance are being established that may differ depending on the place and the culture associated with it. When they are trespassed they are generally not sufficient to land one in prison but they can effectively isolate from the community those that are considered its bearers. Nonetheless, it is important that even conceptually, the freedom of speech represents an essential component for the people's well-being such that, despite the limitations in defining it and promoting it, such freedom merits protection and is considered essential. What is important in this process is what people perceive as essential in the given historical context in which they exist. In other instances, such historical needs involved the ability to participate in the decision-making processes of the state. In ancient Athens, for example, the ability of all Athenian citizens to be part of the political process was an integral component of their well-being, to an extent that probably surpasses the minimal expectations for participation of people even today. It coexisted of course with the fundamental biological needs, but concepts of political freedom and liberty, and active participation in the affairs of the state, were equally important. No life could be full unless it entailed a component of participation in the political decisions.

In both these first two categories of needs, the biological and the historical, all people were entitled to receive them, and the corresponding needs were perceived as essential within the given historical period. The historical period determined the extent of the biological needs that should be satisfied and defined the historical necessities regarding the non-biological needs.

The third category of needs is related to those that are considered privileges and apply to the few only at a given time, but in principle and in theory, they can be accessed by more people in the society. It is the needs that are desired by the people but are not attained by all. Typically, that only the few can have these needs satisfied is interpreted as a matter of supply and demand. Usually, these are the ones that can cause social friction and political turmoil, characterizing the political dynamics of a given society. The satisfaction of these needs is usually reduced to the concept of social mobility that was always inherently present and becomes more and more prominent as we move forward in time. Almost always, provisions were made allowing some individuals to materialize these desired needs and this allowed the needs to gradually convert from desired to historical. The national examination system in China, for example, since 2,000 ago, offered almost everybody, in theory at least, the opportunity to become part of the administrative ruling elite, offering the opportunity for social mobility in a systematic manner. The constraint was that in order to be able to prepare for these exams, aspirants needed the indirect support of their families for their training, and that could be a very limiting factor for peasant families that could not afford to lose manpower in favor of studying. In ancient Greece, privileges due to birth status were abolished during the golden age of Athenian democracy as the opportunity was offered to everybody to serve as ruler, by vote. In parallel, in all cultures, both legends and historical events provided examples in which, because of their merit, abilities, or effort, certain individuals succeeded in accessing more than what had been predetermined by their birth status and heritage. It was the "American dream", discovered by several different cultures long time ago. Buddha came from a rich and aristocratic family but Jesus' father was a carpenter. In another domain, priesthood represented a guaranteed way to salvation, among its other privileges but, in analogy to the national examination system in China, priesthood could only offer this advancement primarily to those that could afford the relevant education, and whose families did not require their individual labor. Good deeds of course were valued for all, but in the absence of indulgences, the effort should be higher, albeit feasible in theory.

Today, especially in the West, the mandate for social mobility, that one should be able to enjoy more than what his or her birth status dictated, is widely accepted to be a fundamental reflection of Western social progress. Or to put it better, one should not be constrained by one's individual heritage. Exactly this is the actual need that deserves satisfaction in the present societies of the West—the opportunity, given sufficient effort and ability, to enjoy a status at which more privileges can become available. And exactly this promise, and its acceptance as such by the people, is what puts societies together and offers them coherence and stability: That the given privileges are not enough for everybody to enjoy since people are essentially more than the available wealth, but everybody could access them if they tried hard enough.

Among these three types of needs, the biological, the historical, and the desired needs, the balance between the historical and the desired is what determines the historical progress. The strict biological needs are stable and easily attainable. They become more extensive over time since the standards of physical well-being elevate, but they remain fairly constant throughout the years. What changes dramatically are the historical needs, which become drastically different over time and location, but are attained for all. The desired needs, however, are the ones that determine the degree of satisfaction of given societies, guide political history and establish the cultural trajectories. From that perspective, unless drastic repositioning occurs in a society, the desired needs become historical and are substituted by new desired needs that will also convert to historical in due time.

Satisfaction of needs and population basis

Over time, assisted by the technological progress during the corresponding historical period, a trend persists according to which what was considered as an essential need, biological or historical, gradually acquired an increasingly wider basis among the population. To that end, population growth was seen as a factor reflecting prosperity, and reduction of the population occurred primarily due to diseases, famine that might follow disruptive periods (due to wars, for example) or climate changes. Population decline was occasionally followed by drastic social reorganization as well. In all cases, however, the individuals' success in the society was not measured by the number of offspring, and that fact disentangled biological performance from social success. This was happening in a more systematic and intense way during the later stages of our cultural evolution, when people started developing civilizations. Happiness and well-being began to acquire meanings that were more and more subjective, and the establishment of historical needs within societies was becoming prominent for their character.

While we see a gradual shift in the balance from the biological to the historical needs, and then to the desired needs, it remains to examine what humanity's biological needs are now and how they become integrated within the whole scheme of needs satisfaction. Biological needs traditionally culminate in reproductive success. A successful individual (and by extrapolation a successful population) is one that has more offspring. Today, however, there seems to be a change in these dynamics. Especially in those parts of Europe that belong to what has traditionally been viewed as the "developed" West, population growth has moved to a negative trajectory. People can certainly afford to raise more offspring than they are producing, and their investment in those they do have does not decrease as the number of offspring increase. Thus it is a conscious decision to have fewer children, or even no children at all, based on other considerations. Perhaps raising them would stand in the way of career plans; and apparently having children no longer constitutes a life goal. Even the commonly posed statement that "people cannot afford to have children because they would not be able to raise them the way they would like to," sounds like an excuse; it indicates that biology is indeed superseded by cultural influences. It is also a reflection of an increased individualism that defeats the biological standpoint, demonstrating that well-being does not necessarily involve biology and reproduction, and one's happiness is not essentially associated with having children and offspring.

During the various historical periods, happiness was determined by how well the historical needs were satisfied, and the more complex and elaborate these needs were (as societies became more complex), the greater the deviation from the traditional biological needs was.

By observing how these needs changed over time, we will recognize a trend. The historical needs cover an increasingly expanding population basis. The desired needs that reflect the privilege of the few at any given time also, over time, are increasingly more widespread, and eventually they too become historical needs. Owning a car for example, the ability to visit the doctor for a check-up, or education, are examples of needs that were felt initially by the few but soon became available to more and more people. Of course, a concept of merit was also present and served/serves as the moral justification for these few to enjoy these benefits before they become available to all, but regardless, it was only the few that could enjoy them and traditionally this type of benefit was considered a measure of the society's standard of living. Furthermore, the collective acceptance of these desired needs, as such, keeps the society together and aligned towards a unified value system in which the same things are aspired to by the vast majority of the people.

As a society progresses and technology advances, historical needs become essential. There are few members of Western societies today who do not have a car or a TV set, do not have some insurance coverage that is either work-sponsored or state-sponsored, are not able to visit a doctor on a relatively regular basis, or do not have access to primary and secondary education. The social mandate evolves such that one can expect these numbers to continue to decline even more, a fact that is reflected in the promises of the political candidates before elections. Those that follow a more left-centric policy will address the challenge by promising state-sponsored programs while the ones that lean towards the free-market economy argue that it will be taken care of by reducing unemployment. Nonetheless, it remains as an unsolved system and it is recognized as such by the society, therefore it warrants policies for its resolution. These are desired needs that have to convert into historical. That these benefits are not yet accessible to everybody is interpreted as a temporary dysfunction for our societies and economic system, something that sooner or later will be addressed and resolved. If, in a given system, personal effort or merit do not warrant unrestricted access to these privileges, it is the society that should satisfy them, by making the education public, and by offering government-sponsored or charity-sponsored insurance plans and care that everybody can have basic access. The support of free clinics through state and private funds and the voluntary work of health-care personnel are initiatives that function towards this direction. As the collective imaginary of the people changes over time, it becomes an intrinsic component of the historical needs to make sure that more and more of the population can meet these needs, even though they are currently classified as desired needs.

That these needs reflect desired social needs and not just demands of special groups of people within society is also reflected in the fact that they engage even people who cannot be benefitted directly. For example, despite the fact that a person has his own employer-sponsored insurance plan, it may be a component of that person's desired needs to live in a society where everybody has access to a basic, state-sponsored health insurance plan. This is true even though providing such a benefit may come at their own individual cost since they will have to cover the others' insurance costs as well. Such behavior is culturally imposed, it cannot be easily explained in biological terms, and it has characterized human history since the earliest stages of our cultural evolution. We have already discussed the care provided to disadvantaged members of the community among Neanderthals.

Here, of course, there is both a catch and a loophole: The catch is that a "historical need" is assigned this classification only once it can be applied to the widest population basis. Before that, it is only a desired need. In other words, accepting as "historical" a need that is not yet unattainable for most people would trigger a crisis. Therefore, it has to remain assigned to the category of desired needs. People usually consider it as an achievement to attain this transition, converting a need from the category of desired to the category of a historical need. Such conversions mark social transformations and the classification can only be done retrospectively.

At the given time that this happens, it is seen as a point of progress. Take, for example, enabling women to vote. It was a desired need as women protested their not being allowed to vote. Soon, it became a historical need, and today in our culture it would appear completely outrageous to argue that women should not vote. Furthermore, a desired need is classified as such only retrospectively, after it eventually converted to a historical need.

A need that currently exists will become desired only if we have very strong confidence that in the future it will become satisfied. For example, global health insurance coverage is a need that is desired because we can state with confidence that in a future society everyone will have direct access to health care and if they don't, it will certainly reflect the system's failure. Furthermore, people travel more and they spend more time away from home. They need health coverage for this. However, having a vacation home, albeit highly desirable by everyone, does not reflect a desired social need because it is not certain that it will eventually be satisfied nor that it can be seen as a historical imperative, at least in the foreseeable future. Nonetheless if, in the future, everyone has a vacation home and that becomes integral to the people's well-being, then retrospectively this will be viewed as a desired need that eventually became historical.

Of course, it has always been debated whether certain needs should become historical, especially for cases that involve distribution of resources such as funds from insurance and pension policies, yet the path forward is set. For a more up-to-date example, access to basic state-sponsored insurance was challenged recently, not on the basis that the economy *per se* could not afford it but rather on the basis that it would harm the economy and the ideals of the liberal society of free choice. Ultimately, it is argued, such policies would compromise the health system by increasing inequalities due to the overwhelming load and pressure in the context of a free market. Yet it is becoming a widely more and more accepted notion, and it seems that likely that it will characterize the future society.

An analogous debate (although it does not develop along financial arguments) relates to the right of women's to abortion. For a fraction of the society, the argument that women have the right to control their bodies and decide if they want to give birth is undisputable, while for another fraction of the society this constitutes a fundamentally immoral and unacceptable act because it takes control over an unborn child's life. This reflects how the cultural context and traditions determine the society's needs. It is plausible that if abortion becomes legal without restriction, it will reflect a currently desired need that was satisfied and became historical. Alternatively, if it does not, for a fraction of the society it will represent a desired need that was not satisfied.

Through this lens, notwithstanding the limitations that are associated with recording matters from our current historical perspective, we can recognize that there is a constant expansion of the essential needs, and that what we recognize as historical progress is the transition of needs from the status of desired to the status of historical. Such historical needs reflect the conversion of needs from non-essential to subjectively essential, while the biological needs can be viewed as objectively essential.

If the reverse happens and needs that were once satisfied no longer are, then this is identified as a step backwards that is only transient and has to be corrected. This happens occasionally following periods of crisis, but their resolution remains live in the collective aspirations of the society. Typically, more people are constantly enjoying more, to whatever level the society demands, and this is seen as progress.

Seeing these transformations from the perspective of material pleasures that become increasingly abundant, the landscape of desired needs is dramatically increasing over time, with several privileges being enjoyed by few, while in the future these same privileges will be enjoyed by more people. Exotic vacations on a regular basis is a privilege of the few, at present; but this will likely become a privilege of more in the future, when maybe the few will travel on a regular basis to space for vacations.

In other cases, different types of needs become increasingly prominent. In certain instances various societies have repositioned themselves and redefined what a historical need is. Take Iran as an example. After the regime change in 1979, the imperial state was replaced by a theocratic state, changing drastically the imposed priorities in peoples' lives. Nevertheless, the ultimate goal was always to offer more people the opportunity to enjoy what was considered to be a need. For example, in societies that abruptly transitioned into a theocratic status, the satisfaction of historical needs was suddenly repositioned, and material needs were substituted by theocratic needs. This was done, though, on the basis that the latter are more important, despite their cost in the former. As such they are accepted, or at least are rationalized by the people, and their continuous acceptance is what will determine the stability of these societies and of the corresponding political systems.

From these perspectives we can recognize in history the process of the constant change in the size of the group that receives the available benefits, as well as the determination and continuous re-evaluation of what these needs are. Therefore, one must study the dynamics that determine how a group is defined from a biological and cultural perspective. In this way the answer to the question between nature and nurture will become clearer.

5. Sociality and biology

Needs contrasting biology, ... or not? Nurture is nature

When I was an undergraduate student in biology, a professor made a striking comment regarding the success of various species. To our great surprise he remarked that microbes were the most successful organisms, not only because they are abundant in numbers but also because they did not have to change drastically over time to adapt to the changing environment. He then asked us to compare the conditions on Earth a million years ago with today's conditions and noted that a winning team should not change. This statement struck me because, like many others in the class, I had the intuitive conviction that we humans were the most successful species because we were the most advanced. From this new perspective, however, humans automatically moved towards the bottom of the ladder of successful species. It is probably a matter of semantics and definitions, but one can readily see the professor's point and understand its rationale. Furthermore, it automatically placed concepts such as success and progress in a completely different perspective.

Progress in humans is inherently linked to culture, and in speaking about human biology and culture, the classical question of nature versus nurture renders the intuitive distinction that the two are discrete entities. Furthermore, it implies that one antagonizes the other, and when one grows the other retreats. Are these concepts truly different, though, in biology? Can humans really live without having a culture being associated with them? Or in other words, could people ever have existed without having developed culture? If we conclude that they could, then what we perceive as culture is not an essential component of our human identity. It is just something that happened, randomly probably, and might very well not have happened. History, though, says otherwise. Not only did culture apparently always develop whenever people came together, living in groups, but also archaic humans like the Neanderthals also developed cultural elements such as burial of their dead and the concept of an afterlife. They used tools, and they took care of their sick and disadvantaged without strict biological benefits being associated with such effort. Apparently, something else beyond their strictly conceived biology instructed them to do so. Traditionally, the distinction between nature and nurture, or culture and biology, was approached philosophically and was reduced to identifying as culture everything that remains when biology is eliminated, and it reaches an apex with the question of whether human nature does exist. As biology advances, though, especially in the direction of biobehavioral sciences it increasingly covers more aspects of the human condition that traditionally were attributed to cultural elements. From that point onwards, the appreciation grows that human nature is broader than we originally thought. However, the definition of human nature remains an open question.

Soon after culture and civilizations developed, and human history began, people began to question the value of culture in their lives—a concern that can only be attributed to humans that are not driven by instinct only. Throughout history, after people conceived their unique position in the world, they started doubting how inherent and desirable culture and civilization are for our well-being. This actually happened at different times, both in the East and the West. Lao-Tze in ancient China was a staunch opponent of the notion that civilization is a requirement for people's happiness. A most extreme version was the position of Diogenes the Cynic, in the Greek classical world. He advocated that people should live like animals, just as they were brought into this world. More recently, along the same lines we have Rousseau's criticism, in the West. They all thought that when people deviated from the natural way of living and invented civilized life, they abandoned their share in living a happy life. In religion as well, there is an analogy in the criticism that developed and in the reference to Adam and Eve, who were expelled from the Garden of Eden as soon as they tasted the forbidden Fruit of Knowledge. Soon after they ate the apple, prompted by Eve, they started feeling shame about their sexuality and displayed traits that were indeed, in their essence, cultural—unique to humans and, among other things, they also developed various taboos. God expelled them from Eden, that is, Paradise, to punish them. At the very least, these stories and opinions reflect a critical stance about culture, knowledge, and civilization, and they question how inherent those are to humans. Such criticism was expressed in both religious and philosophical texts. Obviously, by suggesting that life could be lived without it implies that culture may not be inherent to human nature. Nonetheless, soon after people started living together in groups, they started developing cultural norms, cultures, and occasionally civilizations.

Since the earliest days of their evolution as a species, humans started developing traits that were passed on to the next generation. This transmission of patterns, behaviors, information, and skills, in the forms of memes (in the traditional meaning of the word), constitutes culture, which was obviously very simple at the beginning. Over time, though, it started becoming more elaborate and complex. It also occurred as an independent process, as far as we are able to tell, in different populations of humans and at different places and times, and therefore, an astoundingly high degree of diversity developed. Thus, it constituted an intrinsic component of the human condition, and to that end, it is impossible not to consider it as an integral part of human biology. It recalls the question of *nature vs nurture*, suggesting that this is a false dilemma that can be resolved by the answer that actually *nurture is nature*.

From our own traditional perspective as humans, culture and biology represent distinct concepts; thus, we will continue treating them as separate entities. The distinction will be made on the basis of information that is transmitted from one generation to another, that is either in the form of genes or due to our social interactions with others. Initially, this included information regarding certain skills and tasks via mimesis, or imitation, for example how to light a fire or smash seeds to release edible food. Later, far more information was transmitted, and this entailed inventions for the storage of information such as papyrus, the book you are now reading, or electronic records on how to build a rocket. Besides the availability of information that directly made laborious processes more efficient, it also progressively came to involve patterns of behavior, customs, and social norms. The transmission of information also involved our perceptions regarding our place in the world and in relation to other species, and the afterlife. The details became quite diverse in different human populations. This diversity contributed to, and enforced, the sense of identity of different groups of people, each of which has more similarities in culture than differences.

When we speak of culture today, we mostly refer to this type of information that is associated with our variation in the ways we behave and socially interact with each other, how we define our collective goals, what we value most in life, and how we determine and regulate our interactions with each other. Common culture, in other words, means sharing information and accepting certain things as facts on a basis that is rather subjective. At this point, we have the establishment of the framework that covers what we described before as historical needs.

The scheme we have described so far treats biology and culture as different and distinct domains. The needs people have, were classified as a series of expanding cycles that in the inner position had the strictly defined biological essential needs; then there are the historical or non-essential needs that become essential because of culture: and finally, there are the desired needs. Clearly, culture is tightly associated with the two outer circles which operate in parallel, without one compromising the integrity of the other. For example, the satisfaction of the non-essential historical needs does not interfere with the biological needs of people. And the classification of a need as an essential biological need does not contradict the ongoing attainment of those that have been classified as desired needs, that eventually, strive to become historical.

Exceptions do exist in these general notions, and they happen to become more and more prominent during our cultural evolution as a species, especially in the later stages of our history. Already we mentioned that during periods of social and political reorganization, abruptly, within a single generation, the landscape of the historical needs may change, and needs that were originally considered as essential were abandoned, replaced by others. When the Pilgrims immigrated to America, they were motivated originally by the opportunity to freely practice their faith. To that end they accepted the hardships and the compromise of certain historical needs that could have been attained in their home countries, because they prioritized another need that remained desired in Europe, at least for some people, but had better chances of becoming historical in the New World. On the other hand, when theocratic regimes collapse, they do so, among other reasons, because the collective psyche accepts that the satisfaction of divine mandates is not that essential anymore for the people, and some other mandate becomes more important—usually a more materialistic one. In that case, the latter become the historical needs for these societies.

From a more contemporary perspective, the choice between liberty and equality, both of which represent historical needs of people in the West after the Enlightenment, determine the subtle or the more intense repositioning and description of what the historical needs are for a given society and which ones should be prioritized. Liberty and equality are contrasting notions because to pursue your own liberty, you may need to compromise social equity. Vice versa, the pursuit of equality in a society contradicts the liberty of its members. The active choice of people participating in socialist revolutions, at least at their earlier stages, or even when people in liberal democracies vote consciously for communist parties despite the implications for their quality of life if such a choice were to materialize, represents a prioritization of equality against liberty. And when communist regimes collapse, or when people vote for liberal capitalistic parties against socialist ones, this nominally reflects the prioritization of liberty against equality. In these cases, an impact on the essential needs is usually recorded but this is considered to be a transient compromise that sooner or later will be resolved, because in fact liberty and equality are equally important and neither should compromise the peoples' biological needs. Biological needs such as food or shelter, for example, may be compromised but this is not done purposely. Such compromise is transiently accepted because it happens incidentally as part of a process, regretfully but unavoidably. For example, famine not uncommonly follows revolutions (such as the French Revolution or the October Communist revolution), but that is not by choice, and efforts were made to avert it. Naturally, the extent and intensity of such crises can, in their own right, trigger more instability, causing another repositioning and revisiting of the accepted historical needs, but this is seen as an unintended setback. The famine was never sought as part of the new historical needs as they were conceived.

However, such shifts that may, transiently and unintentionally, compromise the essential needs of people don't always take place within the domains of the historical needs. There are several instances in which the satisfaction of historical needs surpasses the mandates of biology in different societies and historical periods. Actually, this abandonment is considered part of the process. It is a matter of choice on the part of at least some people, and it is considered to be an essential component of a process that aims to satisfy the given historical needs. In these cases, they do not happen collectively, and usually they do not apply to all people in the society. They only apply to a subset of people and may or may not be done by choice for them, but it is accepted by the society as a whole as an intrinsic component of the transformation that will, upon its completion, satisfy the new historical demands and needs. Nonetheless, in all cases the fact remains that for those people their needs are not satisfied; but they should be, sooner than later, and that remains the goal.

In several instances in history, however, there has been a purposeful abandonment of essential biological needs because of cultural interferences. Three of the most apparent examples when biological needs were either abandoned or reduced and deprioritized would be the eunuchs, the (Christian) clergy, and slaves. All have appeared in different versions in different civilizations when social norms dictated that putting aside biological needs is beneficial, and in certain cases, slavery, for example, the moral implications were ignored despite the fact that these biological needs were already accepted as such and were valued in the society. In such cases, ethical concerns were collectively suppressed because the benefits outweighed the moral costs.

In the first case, that of the eunuchs, the right and privilege to reproduce was abandoned in exchange of the opportunity to become part of an elite. Notorious eunuch courts existed in Chinese and Muslim societies because they did not pose the danger of interference with the line of succession. Eunuchs functioned as counselors to the emperors and the sultans, respectively. Frequently they accumulated significant authority and played decisive roles in the politics of the palace. They secured a high quality of life and safety. In that case the abolition of the biological needs was voluntary and operated on the basis that the particular historical needs surpassed the biological ones, since both were considered at this given time as equally significant. The satisfaction of the historical needs that were now essential amounted to a gain sufficient to offset what was lost in biology.

Clergy, Buddhist monks, everyone living the monastic life, offer another example in which needs dictated by historical factors outweigh biology. Offspring production is voluntarily abandoned in favor of dedicating one's life to the divine, for the salvation of one's self or the salvation of the religious community. Occasionally, particularly in the past, joining the clergy also warranted social status and the automatic solution of other fundamental biological concerns, especially in periods and communities of poverty, or during war. The culturally defined metaphysical had again offset the biological.

In both previous cases, the abandonment of the reproductive privileges was a trade off that satisfied the essential biological needs for food, shelter, and protection. In the case of monastic life, especially in asceticism, the trade off also involved abandonment of food and life luxuries to a level below what was considered baseline, or at least historically acceptable, for the given society. People did it because it was thought that it would confer access to divine privileges.

Another historical example, though, where biological needs have been neglected is slavery. As property of their owners, slaves had their biological needs met, including the privilege of reproduction, only at the discretion of their masters. Besides reproduction, other biological needs such as food and safety were satisfied only minimally and arbitrarily, often falling far below the standards considered necessary by the rest of society at that time.

In this case, of course, the big difference is that while the eunuchs and the priesthood were privileged groups integrated into the society, slaves were neither privileged nor integrated. Although slaves constituted a major part of the labor force in some societies, they held no social or political role. More importantly, the term "abandonment" is not applicable since the slaves did not voluntarily give up their other needs. The widespread acceptance of slavery by the corresponding societies stemmed from the fact that although slaves played an instrumental role in providing for the well-being of the others, they as a group were not deemed worthy of protection or the fulfillment of their needs. They were not considered a part of the community. Their "class" was just a tool to be used for the well-being of their owners, who did not have to satisfy them above a level that would keep them functional. The clergy, though, and the priesthood, were integral parts of the society.

Other groups of people with specific attributes, priorities and status, have also defeated biological imperatives. A typical example is the fasting that is common in different religions and traditions. Religious people, at specific time periods throughout the year, may subject themselves to occasionally severe fasting. At that time, they have to defeat their hunger, that may become intense and even life-threatening, in a process that aims, in return, to provide compliance with religious mandates. They are challenged at that time to defeat their biological need for food to satisfy another historical need that prioritizes their metaphysical concerns.

Another example is care for the elderly who, having passed the reproductive age, only represent a cost to the society from a purely biological perspective. Nevertheless, in all societies past and present, albeit with high variation among them, care for the elderly has constituted a fundamental component of efficient social function. Of course, various models can, to some extent, explain it by integrating biology into this. For example, one cares for his or her elderly relatives to establish an example by which he or she will also be cared for later. Another hypothesis may explain it as a strategy by which the elderly are taken care of, because they possess wisdom and accumulated knowledge, and may offer advice when needed. Thus, they have a role in the transmission of information and the evolution of culture. As we will discuss subsequently, menopause in women is related to this cultural attribute and exemplifies the reciprocal dynamics between culture and biology. The wisdom of the elderly will increase the success of the younger that will be able to attain their own needs more efficiently. As such, they also legitimize hierarchical structures and contribute to the coherence of societies. This example appears quite pertinent in Chinese society, as Confucian philosophy attributes a fundamental role to care for the elderly (Mooney, 2016). It also establishes a paradigm where collectively, for the whole society, care for the elderly exemplifies and strengthens a hierarchy, at the top of which is the governing elite, establishing norms that when generalized make the society resistant to change.

Various acts of valor through which some individuals voluntarily give their lives for the benefit of others, are common—and indeed essential—for the establishment of a collective psyche, in all societies that have their heroes. These also provide well-known examples in which the pursuit of the essential needs (food, shelter, and protection) are not attended to; rather the aim is the attainment of other goods.

Of course kin selection, especially in cases where the distinction between different groups has some genetic basis, may partially explain this. For example, the revolt of one ethnic group against another may have a biological foundation since it can be seen ultimately as benefitting the genetic relatives of one's compatriots. Nevertheless, its impact surpasses biology and in most of cases genetic similarity is substituted by cultural similarity, which is more vague and less strict. Cultural similarity also allows integration of additional people in the group, facilitating its rapid expansion. Actions of religious fundamentalism can also be relevant to this, in some remote sense, in view of the fact that the individual may sacrifice his life as a consequence of metaphysical demands. Religious fanatic suicide bombers throughout the world, or the Japanese kamikaze pilots in World War II, elect to die in exchange for accomplishing their greater purpose, which their culture and ideals mandated. Political motives, unrelated to ethnic or religious priorities are also present throughout history and provide similar examples of people that choose to sacrifice themselves to improve the lives of others.

All these cases in which, from the perspective of genetic relevance, the biological needs are substituted by cultural mandates, can be attributed to the centrality that the historical needs acquire, and as part of the process through which the desired needs will eventually become historical. When a priest decides not to have children and instead to dedicate his efforts to the salvation of others, he does so because the society, and he himself as its member, have identified the desired need for salvation; he acts accordingly so that salvation may become a historical need. The operation of the concentrically expanding cycles of the essential biological towards the non-essential historical, and then towards the desired needs, is still valid because as the value of culture is raised, it becomes easier for people to reduce the significance of biology.

Beyond the specific groups of people that follow these specific norms, the whole technological civilization we currently live in can be seen as a compromise of our biological needs. By accepting that our lives are lived in a manner that is less healthy and less "natural" than it used to be (because of the food we consume or the destruction of the environment that we collectively, both as producers and as consumers, are engaged in), we automatically acknowledge that a shift in the historical needs has occurred, such that biology is compromised. On top of that is also the fact that we have far fewer children than we could rear with our existing resources.

Of course, this now touches again on the question of whether "artificial", as opposed to natural, is only something that is new, and over time it will become natural as well. In other words of whether we define as artificial and as product of culture, something, only historically, by comparing it to the past. We feel that something is currently part of our human nature, but we can appreciate that it is the product of our culture by thinking that in the past, or elsewhere maybe, was not perceived as such. Take religious beliefs as an example. It is widely thought that the belief in a higher entity is a need inherent to human existence because it has been present since the earlier stages of our history. Thus, it is part of human nature. Nonetheless, especially in the last decades, increasing numbers of people confidently abolish such a belief, implying that the religious belief is just a product of our culture; and conversely, even the abolition of such belief is a product of our culture as well. Wanting more material goods and having higher spending power is another example. None will refuse higher income as long as it is not associated with some personal cost or sacrifice and even if it is not associated with wealth that can be used by the individual or the ones they care about. Thus, it can be interpreted as part of human nature. Yet this is something that is applicable especially for our current cultural and historical track because at earlier stages of our history, the accumulation of wealth was not as central as it is now.

It does not really matter, though, because it all applies to the sphere of historical needs; and whether they deviate just a little or more from the natural as defined biologically is only a matter of terminology since, strictly speaking, everything is natural just because it does happen. People in their everyday lives do not have the ability to distinguish between what is mandated by historical factors and what is really inherent, and they live unable to separate the two. Choices, feelings, and position in society are all based on the assumption that biological and historical needs are equally important and inseparable. This is why it is easy to intermingle the two, although when their boundaries are evaluated historically, from an academic perspective at least, they appear to be clearer. For a person living in a tribe in the Amazon, or an elderly lady in the Middle East who needs to attend the Mass every Sunday, the attention to rituals appears to be subjective only when they are studied by cultural anthropologists. For the individuals in question, they are as objective and imperative as they can be.

History as the change in needs rankings

Human history is rich in examples where biological and essential needs are replaced by the historical needs that may have opposed them. Over time the latter change in character and intensity, and they were never viewed during their contemporary times under the relativistic perspective that we recognize today when we study them. In their times, the historical needs were perceived as equally important, and frequently even more important, than the biological. That being said, in prehistorical and early historical times the biologically essential needs were more dominant, while concomitantly with the emergence of civilizations, especially at their peaks, the historical mandates that shaped these needs appeared to become more prominent. History seems to favor the needs that are dictated by culture.

As the civilizations matured further, though, approaching their collapse, these historical needs appeared to lose their prominence. This probably reflects the evidence for social dysfunction and the deterioration of the civilization. When the historical needs that derive their value from ideals and abstract social concepts start to become unappealing, people seek to replace them with other historical needs, but in the meanwhile, biological needs seem to predominate. Following wars that alter the collective mind sets and aspirations, the quest for survival replaces other needs. People abandon the heroism that may have inspired them before to exchange their lives for the benefit of others, and now they just want to survive by whatever means necessary. In these cases, and in these periods, they reprioritize their biological mandates, abandoning the abstract historical needs. The collective ideal of a happiness that was dictated by cultural norms loses its prominence when the biological needs are not attended to. People in these cases become more sensitive and receptive to new cultural trajectories that can reposition them and make them adopt new historical needs. The fall of the classical world in Greece was associated with a prolonged period during which hedonism acquired prominence; this was expressed by the Epicureans. This in turn provided a foundation for the spread of Christianity thereafter, and that, among other things, also oriented peoples' needs towards more spiritualistic priorities.

From an opposite perspective, when cultures were developing and civilizations were emerging, biological needs appeared to be less dominant. At these times people, moved by collective ideals, compromised their biological needs in order to achieve the new collective goal.

The onset of a civilization is synonymous with the strengthening of a collective imaginary in which biology becomes secondary to culture, or at least it becomes equivalent. Before the collapse of civilizations and after their peak periods, cultures are positioned second. When civilizations are emerging and collective cultures are developing, the voluntary sacrifice of individuals for a collective good is viewed as noble. It is also noble and widely acceptable in the society to live lives within niches where biology is undermined, and the pursuit of one's goal in life may supersede biological well-being. This could be exemplified by entrance into the priesthood, being an artist or a scientist, or even going into politics from a romantic standpoint, when people do not prioritize their personal well-being over the call for a mission they conceive in life. The existence of such individuals offers societies coherence, provides its role models, and strengthens the significance of the historical needs. By prioritizing the historical over the biological needs, people establish systems of principles, the validity of which is subjective and understood by them, the people of the same group, which is equivalent to the establishment of cultures and civilizations. This is the condition that makes them choose to abandon their biological needs in order to attain those that are historical.

This may lead to the rather pessimistic realization that the Western civilization as we experience it today is headed for collapse because the attainment of the peoples' individual needs emerges as its predominant feature. Furthermore, it is illustrated by the fact that the collective culture becomes looser as it strives to accommodate as many people as possible. Of course, these needs that are perceived as biological but in their essence are historical now are much broader than ever before and involve a wide spectrum of needs that target the materialistic aspects of our current culture. The West has entered a stage in which practicalities have become more important than the ideals of past periods. In this case, the cycle of the historical needs is expanding, but it only relates to how biological needs are interpreted. For example, a comfortable, fast, and safe car is seen as a pure extrapolation of our biological imperatives that allow us to move more efficiently from one place to another, indulging our senses and reducing our discomfort. A huge effort to make medical progress is also directed at extending our biological capabilities and extending our reproductive capacities against age restrictions or disease limitations, even if we have consciously chosen not to exercise them, in terms of our reproductive success and the efficiency in our contemporary societies of the West. We want to have the choice, though, and we actively invest on that even when, as a society we do not exercise that choice.

Groups and group membership

Throughout human history the collective good was always central in driving change and progress. This notion applied in genetics as well, when altruistic behavior emerged and guided evolutionary processes. The term "collective," though, is rather ambiguous and implies the existence of a group that will receive the benefits while those outside the group will not. After all, there must be some type of competition in order to attribute value to notions such as altruism, sacrifice, and benefit. Otherwise, if it is impossible to establish who the other is, they become meaningless and ambiguous. Thus, understanding how the group is defined is essential.

The definition of the group remains among the most controversial topics in human history, not only because of the opacities in its definition but also because of its implications and the realization that it has guided and still does guide major conflicts. Aristotle famously described humans as political animals, meaning that peoples' natural context is the Polis, and actually a fairly small one, the size of which should not compromise the ability of one to participate in the ongoing—political—affairs. From a different perspective Lao Tze, too, on the other side of the ancient world, saw happiness attainable at small villages, where the complexity of administration is minimal and does not subject individuals to the hierarchical authorities that are foreign to the peoples' true nature. Famously he described happiness as living in a small village that is close enough to another village that you can hear the roosters crowing. Nonetheless, you don't ever need to go to the nearby village because you have everything you need to be happy in your own village. Confucius was also a strong supporter of the essentiality of a group, of the society, but contrary to the others he had the Chinese kingdom as his prototype and the target of his philosophy, the success of which was intrinsically related to the well-being of the Chinese society. According to Confucius, the supporter of the centrality of the social relationships as the main constituent of peoples' well-being, it is our relationship to others that determines our happiness, and only if such relationships are appropriate and well-functioning we can live happily and purposely within the society. Yet, both recognized that living in a group is the ideal environment for human beings. During the most recent phases of European history, the group that should operate as the unit within which each person should identify is their nation. At the same time, Marx identified the class as the major unit with which one should identify and align.

So far, we are talking about people in relationship to a group that they are part of. Within this group they try to accomplish their goals, and their successes are usually defined in relationship to this group. There is both a biological and a cultural rationale behind this. Group membership is favored by people throughout the history of our species, on the basis of the fact that the benefits outweigh the costs. And there are costs in group membership. More importantly you need to allocate a portion of your personal effort and resources for the benefit of others, and occasionally, even give your life for them. You also need to act following the choices that others have made, compromising your own will and opinions. These are all costs that are justifiable because they provide advantages, or at least ensure the persistent satisfaction of the essential biological needs such as food, shelter, and protection. In turn, these guaranty the privilege of people to reach reproductive age and produce offspring that, by their subsequent integration into the group, will receive similar benefits for them and for their offspring. This at least reflects the biological perspective according to which genes, individuals, populations, or species must operate in a way that ensures their persistence. To that end, our own individual compromises come to the benefit of our genetic relatives, our family for example, in order to make sure that our shared genes will persist. I may choose to die, in a battle for example, to increase my relatives' chances to live.

The group is, or at least can be, quite specific in biological terms, and depending on its definition, the goals and trajectory can be speculated. To that end we can assess, even mathematically, whether a cost is considered to be reasonable biologically, based on the degree by which genetic relatives are benefitted and how many of those do so. Nevertheless, as soon as cultural constituents are engaged, things become vaguer. What defines the group for those purposes remains a notion that is subjectively defined over time. In the communities of the earliest periods of humanity, in prehistoric times, it was bands of individuals that were related by family bonds. In these cases, the genetic resemblance underscores the group's constitution and recalls the operation of similar strategies in animals that have been exploited on the premise of kin, and group selection. So far, biological attributes are clear, and their impact can be appreciated by a manner that is fairly simple and direct.

The concept of the group, though, has expanded over time and started to include increasingly larger number of people. During these processes the biological component begins to lose its significance and cultural attributes become more prominent. Bands gave way to tribes and eventually these led to nations and even races. The term "race," of course, is also rather ambiguous and unjustified scientifically, but here it is used as an indication of higher genetic similarity between different populations. In order for someone to identify himself in such larger groups, the impact of cultural attributes becomes increasingly important and includes language, traditions, customs, and everything that defines someone's identity in terms of traditions and history. It also includes the acceptance of someone else's authority—the power to declare wars against others, to impose taxes, to pass laws and regulate financial transactions, to use force and put someone into jail—but who also has the obligation to defend their subjects or citizens and support them.

We can recognize again that over time, especially in the phases when cultures become more coherent, the biological imperatives retreat and give way to historically defined attributes. This is the time when the society is considered to be functioning well, and the turmoil is diminished. When people started establishing discrete and coherent cultures that in turn transformed into civilizations and their states expanded, an astoundingly high degree of genetic diversity was incorporated into their groups, where cultural attributes provided the cohesive bonds that brought people together and allowed them to identify norms and ideals that were common. This was far more apparent when civilizations were associated with the establishment of empires which had to be able to integrate different populations and their distinctive cultures.

In most of these cases, within multicultural empires, one culture is dominant. This ability to incorporate different cultures, carve them accordingly, and assimilate them, is essential, and it relies on the plasticity of the cultures and their adaptability. In other cases, a group of people that carried distinct cultures has retained its identity despite being part of a larger state with a different culture. For example, both Jews and Palestinians have lived in a certain region since ancient times, as part of various empires, kingdoms or states, but never assimilated to a significant degree. They only recently established states in which their cultures and traditions can be dominant. And yet, as these lines are being written, the integrity of these states is compromised, while at the same time the corresponding people feel their sense of cultural identity intensified.

But regardless of the specific details of how all different populations were organized within the larger group, the expectation remained the same: All people, regardless of whether some benefitted more because of their cultural traditions or social status, expected to have at least their biological needs met. If there was oppression and those needs were not attended to, then crises emerged, and a progressive delegation of power occurred that was linked to certain privileges. If this did not happen, usually violent conflicts emerged and that was followed by the dismantling of the associated groups, a process that eventually would deliver other groups among which the benefits are distributed.

Not only the biological needs but historical needs had to be satisfied. Besides prosperity and quality of life, as they were perceived in the corresponding historical times, cultural needs were also important. The large and successful empires of the past, such as the Roman empire, the Ottoman empire, or even the short-lived Mongol empire, all had in common their diversity as well as their resilience and tolerance for different cultures. The fact that they could persist was exactly due to their ability to allow different cultures to operate and function in parallel, and as such, the historical needs of the corresponding people could be satisfied.

This view bears some analogy with the Hegelian perspective on history according to which its ultimate goal is to reach the freedom of the human spirit through a process that is dialectical and proceeds continuously via the synthesis of a thesis and an antithesis. In the cases we have described, it is the culture that determines how such freedom is perceived in any given period. Nonetheless this remains a rather subjective perception, and therefore a linearity cannot be easily recognized. Such linearity can only be seen if it is evaluated exclusively by the standards that are valid for a given culture. Outside these standards, it may acquire different and frequently controversial meaning. Freedom, instead of its typical meaning as we interpret it today in the Western world, should be rather seen more generally as the relief from any form of oppression according to which the historical needs of a given period are being suppressed, and the freedom is to materialize the desired needs. In various instances, such needs may oppose our current understanding of freedom, but they do reflect the decision to satisfy the new historical needs. For example, when pre-agricultural societies committed to agriculture and the hierarchies associated with them, freedom as it is widely perceived now, was abolished. But it did reflect a relief from the oppression that was imposed by the lack of sufficient resources and the uncertainty that characterized those societies. From this perspective, foragers became free despite the fact that they started living in hierarchical societies.

How was group membership decided? Throughout history, people were never asked whether they wanted to belong to a group and what this group would be. It always came as a birthright. Expectations in life were linked to where you were born, and to whom. It also determined who your enemies were and what your desired needs should be. The same also applies to the scale of the group. One could identify himself primarily as a Frenchman, as a member of a specific tribe in the Amazon, as a member of the communist party of Italy, or as a member of the Church of Latter Day Saints in the U.S. This identity defines one's over-arching mindset and goals, and a significant component of their historical needs. Second to that are different group memberships of larger or smaller scale, but usually there is only one defining group and it is subjectively perceived. A member of the Church of Latter Day Saints in the U.S. will find it easier to connect with a member of the Church of Latter Day Saints in Brazil than another American who is Muslim, and a communist in Italy with a communist in Greece than a right-wing Italian, as long as they feel this is their primary cultural identity.

There have always been tweaks in this, and the ability to mobilize oneself within the larger group varied. But if, for example, you were born French in 1870, you knew that your enemies were the Prussian, or if you were born Chinese in 1300, you knew that you had to live under Mongolian rule and accept that a different group of people (again, not by choice), could exert power over you. Furthermore, these days, if you live in Western Europe you expect that your government will defend your property, will offer you a pension when you retire and will take care of your medical bills if you get sick. In the US, on the other hand, you understand that maybe the former will not be taken care of as such, but at least the government will provide you the fair opportunity to take care of all these yourself, as long as you are healthy and able to work. No one asked you in advance, but you live up to these anticipations that come with the way the society you were born into has been organized. From that point on, people live in accordance with these norms and anticipations, and depending on the flexibility provided, they try to do the best they can in terms of fulfilling their goals and satisfying their historical needs.

There is a margin in defining the group's priorities that expands or retreats according to the period that formulated the desired needs. For example, in China in the 14th century it was becoming increasingly evident that eventually the Mongolian rule might end, as it did. A desired need was building up at this point for the Chinese to liberate themselves from the Mongol rule. Similar to this, in the U.S. an increasing demand is developing for an administration that makes provisions for minimal healthcare and guaranteed pension, on top of the fair opportunity to provide these for oneself. This also reflects a historical need that has not yet been satisfied, and therefore it reflects a point of friction within the society. It comes as a criticism of the current social norms, and the fact that it has not been attained yet is only due to financial and logistical reasons. The failure to attain it is not based principally on the "unfairness" of such demands.

The criticism regarding the French people's perception of Prussians as enemies in 1870 is analogous. It was a fact, but that did not mean that this attitude had to remain as such over time. It seems that we enter life and have our group membership established not only as biological beings with a given genetic heritage but also charged with a very specific cultural heritage as well, one that defines who we are and how we have been developed collectively by our ancestors. Then we need to function within the borders set by our biological and cultural context. There is, however, a certain degree of flexibility at the personal level in redefining our needs and aligning ourselves with others that share similar ideas with us, within the larger group.

While this is valid for most cases, it is not always true that group membership comes as a feature determined at birth. There are also instances when group membership is voluntary. This is fairly common in the domain of religion, where people may consciously abandon the privileges and mandates of the group they were born into and choose to identify themselves with another, occasionally with conflicting priorities, to satisfy needs that belong to a different sphere. The divine, for example, the quest to live a life that possesses higher meaning, and the expectation of an afterlife. In that case, a different set of needs may become primal, but again, one's identification with this group is what will decide the course of one's personal history. Whether this choice benefitted others, such as in the case of clergy and others who dedicated their lives to helping others from a religious standpoint, or in the case of asceticism that only impacts the ascetic but does not others, is irrelevant. In both cases the call applies to the individual that makes the choice to abandon the values that predominate in his or her society and prioritize other values.

The United States is another example of voluntary participation in a group. As a country it was based precisely on principles established by people that originally, of their free will, decided to form a nation founded on a series of commonly accepted ideals. These ideals, being desired needs that could become historical in the new country, spanned religious beliefs and work opportunities. One way or another, they established a foundation that persisted. From that point onwards, people massively immigrated and continue to do so, precisely because they share these values and have consciously decided to live their lives accordingly. In that case they consciously decided to engage in a society that attains different historical needs, which may be very similar or very different to the ones of the society and the group they originated from.

In other instances, people join groups that are not necessarily defined by political organizations and states, but rather cultural movements or subgroups within their established group. As these smaller groups acquire increasing prominence in peoples' lives they may come to dictate their norms, priorities, and their whole imaginary. The hippies, for example, in the 70s, or mainstream political left wing political movements of the previous century and today, did not recognize the state as the major power that determined group membership. The group in these cases transcended states and borders and recognized commonalities with people in different regions and countries who identified with these groups. Adherence to religious groups that operate within the liberal democracies of the West also provides benefits through the satisfaction of needs that people may identify as historical. It is the laws of the state that allow and protect the operation of these groups by fulfilling the demand for religious tolerance, and the groups themselves are responsible for attending to the historical needs, fulfilling the peoples' metaphysical concerns.

The pattern again remains the same. People start life carrying the burden of the cultural heritage that is already assigned to them and then, within the margins defined by the corresponding society, they may choose to align within a subgroup that may serve better their perceived historical needs. Such alignment guides the decisions and expectations of the whole society and the ruling class, the administration of the group, and its ideals.

At a more philosophical level, when the political organization is adopted on a voluntary basis it relies on a social contract. This was advocated by Hobbes, Locke, and Rousseau, according to whom all government originates from the people, exactly with the anticipation of satisfying the peoples' needs and demands. When it stops doing so, it is an unworthy government that should be changed. The expectations are generally similar to those cases when the government does not rely on a social contract. Again, it is the administration's obligation to see to the well-being of people, whether this derives from a mandate from Heaven, as was underscored in the Chinese dynasties, or it could represent an unwritten obligation that offers the rulers legitimacy. They all had to prioritize the peoples' well-being irrespective of how this well-being was described and perceived. The idea is also applicable when the subgroup in question is not of the magnitude of the state but expresses a distinct set of ideals that can co-exist with those of the society overall. For example, one's perceived historical needs may prioritize metaphysical questions and a role in helping others. In that case, which corresponds to voluntary group membership today, the individual may join the priesthood. If for any reason this choice is not deemed adequate in fulfilling their goals, they may choose to dissociate from the priesthood and live a secular life, or even establish a new priesthood or religious movement, with the hope that it may be appealing enough to help others attain their historical needs. The Protestant Reformation is exactly such an example; it grew out of a perception, on the part of a significant number of people, that the Catholic Church had stopped fulfilling its fundamental mission and a new religious movement had to be initiated. All these transitions relate to the sphere of people's historical needs that occasionally have absorbed or minimized the impact of biological needs, which might be either fulfilled or eliminated. In all cases people, depending on what priorities they have established, may join a group that aims to satisfy their historical needs.

Why do people want to live in groups, and why do they tend to become larger?

That people prefer living in groups is well recognized and has been documented throughout our history. There is also a trend for the groups to become larger over time. Bands and tribes eventually were replaced by smaller and then larger kingdoms, then empires, and more recently by unified large zones of high cultural homogeneity within which—despite the borders between them—the similarities in their collective ideals are greater than their differences. Some may argue here that cycles of expansions and contractions is what is really happening, and people do not really choose larger groups. Larger groups do occur because of the dominance of some against others and eventually they disintegrate, and this goes on. Nonetheless, "group" here is not identical to "state" but rather to cultures. For example when large empires or kingdoms break down into smaller states, the culture that defines the group does not necessarily break down as well and disintegrate into new ones. Upon the collapse of the Soviet Union, an artificial conglomerate of different cultures, the resulting new states have attended the common historical need for all to function as independent nation states. This reflected their actual desired need before, that was suppressed by the attempt of the Soviet regime to attain the historical need of establishing a communist state. Thus, the existence of several different nation states fulfills the conditions of being part of a larger group that retains national identity, operate as liberal democracies and eventually will join the even larger group of Eurasia.

Larger groups disproportionately benefit those in power, who exercise their dominance to expand the size of the group and increase their quota. This applies as well to states with more power within the group, or to individuals that belong to the higher class within the state. This is true, but in that case, if cohesion is just maintained by power and force, sooner or later the whole structure will collapse. Those who are not in power have to be integrated and they have to receive benefits that surpass the costs. This notion is exactly what constitutes the foundation of a construct, be it a state or a multistate organization with similar ideals, that has to support the ones in need. In the first legal document known to humanity, the Code of the Babylonian King Hammurabi dating to around 1750 BCE, it is explicitly stated that his mission was "to prevent the strong from oppressing the weak" (Roth et al., 1997). This recognizes that a ruler's main task is the protection of the weakest and the assurance that such benefit will be received.

Humans are hard wired for imitation

To appreciate the dynamics that favor a group's expansion, we first need to understand why humans became such a successful species that they spread around the world and inhabited diverse environments. We will see that this success is largely because of our ability and inclination to develop cultures that guide our ways by means that eventually extend beyond our biological limits. And to do that, and to leverage this ability and exploit our potential, we need to live in groups. When the groups are larger, our chances for success increase as well.

We are not the fastest nor the stronger species. We are not also able to eat food that is not cooked, and we are not able, like other animals, to intuitively distinguish which seeds or plants are poisonous and which ones are not. We rely to others to teach us all that information, others who received the information from others, and they from others, that ultimately learned it the hard way—by experience. The human reproductive program—the prolonged period of gestation needed to deliver a baby, and the fact that we have to tend our offspring for some considerable time after birth, makes the burden of care for others besides ourselves even heavier. Not only do we need to provide food for our children, we have to teach them how to become self-sufficient both in terms of protection and in terms of getting their own food. Despite the considerable effort to do this, humans like to explore. We have spread around the world and occupied environments that are drastically different from each other without bearing any significant genetic adaptations for that (despite the public perception that northerners, for example, can do better in colder climates while southerners can do better in warmer ones). The biological truth is that we can all, more or less at least, do just as well or as poorly, in these environments. Northerners do feel cold in the north and southerners do feel the heat of the tropics to similar extents. They have just learned to deal with them. All each one needs is the appropriate equipment, clothing, and some knowledge regarding how to use them and how to cope in extreme weather. With that knowledge, anyone can do it, even the weaker of us.

There are two main possibilities that can explain that to a sufficient degree. According to the first, it is our intelligence that allows as to improvise when we deal with hardships. According to the second, it is our ability to learn from others, meaning that we do not always have to rediscover this knowledge from scratch, every generation. Numerous studies from the fields of psychology, evolutionary biology, anthropology, and other related disciplines have addressed the question and concluded unanimously that it is not our intellect that is responsible. The reason is very simple to understand. We don't have to discover everything ourselves.

A series of experimental studies in toddlers and chimpanzees was very illuminating because they highlighted the differences that may provide the foundation for our cultural evolution. Toddlers were involved in the study not only because at that younger stage intelligence is not that advanced yet, but also because there is less interference from cultural norms. Therefore, the responses that were recorded could be attributed to our intuitive inclinations. An array of such studies failed to demonstrate that toddlers are capable of performing substantially better than chimpanzees in a series of games that address the ability to learn the simple scissors-paper-rock game (Gao et al., 2018), or regarding the sense of fairness they possess (Proctor et al., 2013a, 2013b). Apparently, both species can learn equally well, and also, their intuitive morality is not very different. However, when the game requires cooperation in order to succeed, especially under risky conditions, children perform better (Duguid et al., 2014). Such experiments are important because by comparing our behavior with that of a closely related primate, conclusions regarding the evolutionary origins of our behavior can be obtained, and we can appreciate why only humans have developed cultures. In such studies, what seems to make the difference is not the cognitive skills for the understanding of the physical world but rather our social skills and how they can integrate information passed on by others, for our goals (Herrmann et al., 2007).

Furthermore, the average human is also uniquely able, unlike chimpanzees and children with autism, to exhibit shared intentionality, to share goals and act accordingly to attain them (Tomasello et al., 2005; Tomasello, 2018). Thus, living in a group is not simply a preference, just because we feel protected within it and safe. Apparently, humans have the unique ability to leverage the information that is provided by the group, align their own goals with the goals of others, and maximize their individual benefits. This appears to be a part of human nature and is related to the development of a cumulative culture, something that appears to be distinctive for humans (Henrich, 2015; Tennie et al., 2019), although hints at its existence in chimpanzees, as well, have been suggested (Whiten, 2017).

Of course, the maximization of one's benefits, in absolute terms, is not always the driver of the various choices we make, and the presence of a partner does not always make a positive impact. In another game, involving children and chimpanzees, a substantial difference was recorded when the payoff was evaluated in comparison to the partner's payoff. In that case, it is the relative rather than the absolute payoff that mattered. In this experimental study, older children 9–10 years old, selected to reduce their partner's payoff even if that came at a cost to their own payoff (Herrmann et al., 2019). Younger children, 5–6 years old, and chimpanzees however, selected to prioritize their absolute payoff. Thus, for the older children it doesn't matter if one gets less, for as long as the partner will get even less. Such a choice, which may seem irrational, is dictated by a social comparison and underscores the impact of human sociality on decisions that are not necessarily always beneficial in absolute terms. Again, it becomes apparent that the dominance of the group determines our success in relation to the loss of the others. Already our choices become collectivistic, albeit in a negative competitive manner, rather than individualistic.

A famous experiment performed almost 20 years ago compared children and chimpanzees in their ability to learn by imitation, on how to open a box and retrieve a reward (Horner and Whiten, 2005). The catch was that an irrelevant step, using a stick to tap the box, was included. The tapping did not influence the retrieval of the reward. The study showed that children, but not chimpanzees, would open the box and irrationally also include the box-tapping in their process. Chimpanzees used their hands, eliminating the irrelevant tapping step. What is probably more surprisingly is that older children and adults irrationally included the box-tapping step even more meticulously than the younger children, suggesting not only that imitation is in our nature but also that as we grow older, despite our likely belief, we do it even more rigorously. Tapping the box could be perceived in that case as the cultural element while the other steps reflected the instinctive attempt to receive the reward. This observation provides the core for the cultural diversification of people because it shows that we have the tendency to follow what we see, quite meticulously sometimes, without questioning its foundation and potential impact or benefit, and although we occasionally try to interpret it as cognitively meaningful (Lyons et al., 2007). We do it just because others do it, like a habit, and we expect others to do it as well, establishing a tradition. These traditions acquire the power of giving identity to the ones that follow it, and they emerge as cultures that eventually and occasionally may also progress into civilizations. It is exactly this fact—that subjective truths are accepted and pursued as valid—that signifies culture.

Imitation, which is profoundly advanced in humans, is the instrumental process that enables learning and acquisition of skills. The power of mimesis in assisting us to adjust to unfavorable environments is well known in real life and has also been documented in various historical and anthropological studies. In one such example, explorers in the 18th century were forced to live in the arctic zone. Those that remained isolated and did not manage to intermingle with the local populations and learn from them how to deal with the adverse conditions, how to obtain and prepare food, and how to protect themselves from the cold weather, died. However, those that had locals with them to teach them what to do, succeeded (Henrich, 2015). Thus, cultural assimilation through mimesis offers a vital advantage related to biology and survival.

Mimicking is a component of human societies, and in social networks it has been demonstrated that habits, even diseases, tend to follow network structure (Fowler and Christakis, 2009). Thus, through behavioral traits that may have or not a genetic component but are transmitted culturally, by mimesis, subnetworks are developed that supply the elements of diversification of human societies.

There is another twist in the dynamics of imitation that adds to the complexity the societies of humans acquire over time, and this also explains why larger groups are more successful than the smaller ones. When children imitate others, they don't do it randomly, in an unbiased manner. Preference is given to those whom they identify with and connect with. Earlier studies have shown a preference for own-gender models such as parents, teachers, or peers (Basow and Howe, 1980; Gilbert et al., 1983; Slaby and Frey, 1975). This type of same-gender imitation was also more prominently associated with the activation of brain regions that have been linked to processing reward and reinforcement learning (Losin et al., 2012). Thus, an interesting type of dynamics emerges that may assist in understanding why groups have the tendency to become larger over time, and why the larger ones, despite their problems and potential sources of instability due to their size, are more efficient altogether. In a larger group, the pool of individuals that one can imitate is larger, and therefore the chances are higher that skills that are collectively more efficient and beneficial will be identified and stabilized within the group. This eventually takes on dual significance: It not only identifies and selects through trial and error certain patterns and norms that have beneficial consequences, it also stabilizes behavioral patterns that contribute to identity and the coherence of cultures. Consider for example one group of 20 and two groups of 10 individuals. If, in the group of 20, an individual comes up with a better strategy to make a tool or comes up with a useful strategy for the preparation of food, this will spread more readily among the 20 individuals of the first group. In the 10-member group, the strategy has to be discovered two times independently, once in each group, before it spreads to the remaining 9 members. In the 20-member group it needs to be discovered just once and then it spreads to the remaining 19. Thus, being a member of the larger group, even if you are not the one that came up with the invention in the first place, increases your chances of receiving the benefits of this development. Progress and advancement appear to correlate positively with the size of the group, as they spread more effectively; and that in itself attracts more members into the group to share the benefits. This also benefits the ones that came up with the invention, because having more individuals that can potentially imitate them raises their status in both absolute and relative terms. And this attainment of status is also a goal that is associated with biological benefits such as reproductive success and better access to resources.

This leads to the collective-brain hypothesis which proposes that innovation is achieved more efficiently in larger groups and surpasses the ability of the specific exceptional individuals that may be present in the group. An interesting analysis evaluated the degree of optimization of words and the number of speakers of the corresponding languages (Muthukrishna & Henrich, 2016). Word optimization reflects the information content of a given word, such that a highly optimized language has small words when they carry low information values and vice versa, correlating word length and information content. It was shown that when a language is spoken by more people it becomes more optimized, producing the outcome that would be delivered if a collective brain had undertaken this optimization task. In this sense, being a member of a larger group makes one smarter from the perspective that it enables one to communicate more efficiently.

Coming back to our description of needs from an earlier chapter, the prospect of an increase in status may very well operate as the foundation of a desired need, which may encourage us to act in ways that we anticipate could help us become one of those that receive maximal benefits in a given context.

Of course, as the groups become larger and the societies become more complex, the diversity of those providing models for prospective imitation by the copiers also increases. To address these points and facilitate the selection of the most beneficial strategy among those that become available, additional mechanisms have evolved that implicate prestige as a determining component of social function. This was shown to operate distinctly from dominance for the attainment of status (Henrich and Gil-White, 2001). Thus, an evolutionary trajectory was established according to which individuals resorted to increasing the prestige they possessed in the society, thus raising their status and maximizing benefits. Through this mechanism they functioned as models for imitation. This was a more efficient strategy than attaining status through dominance, which could eventually be detrimental for a society. Competition, in that case, progressively converted from competing in force to competing in ingenuity.

Social learning and culture after agricultural

Projecting these dynamics to the human societies, we see that it becomes natural for a group to continue expanding for as long as the resources permit it, until the point where such expansion is compromised by internal instability or when there is a conflict with another group that is also expanding. Expansion also stops when natural disasters or diseases drastically impact the population. Such events have happened frequently in human history. More recently, the Black Death, a bubonic plague pandemic in Europe in the 14th century, caused the death of about 50 million people or about 50% of the population. This detrimental event paused the population growth in the agricultural feudal societies of the Middle Ages and operated as a catalyst for the changes that soon thereafter resulted in the Renaissance.

In small-scale primitive societies of foragers, the limited resources were the decisive factor keeping populations density down and group size small. Evidence obtained from the examination of 21 forager societies suggests that generalized warfare remained restricted, with homicides representing the most common form of lethal aggression (Fry and Söderberg, 2013). Clashes between groups did not constitute a major factor in limiting their expansion, with resource availability being the major factor for that.

These dynamics changed after the emergence of agriculture, which was associated with the considerable scaling up of the societies, their increased complexity, and of course, their immobility. In these societies, where the population density was increasing drastically, imitation took on a greater weight in maintaining and promoting progress and played a prominent role in maintaining their over-arching cultural themes as well. More elaborate techniques were now demanded for food preparation, storage, and distribution, and the spread of these innovations had a population base that was considerably higher as well. Groups were also becoming larger, both in order to sustain the progress and because agricultural societies were immobile settlements. The population that increased due to the surplus stayed around the regions that produced it. In turn, in order to sustain themselves as such, the societies needed to function on the basis of cultural norms that were more stringent and tighter, a fact that was both a cause and a consequence of the demands of the increased group sizes. Societies that are culturally tighter enable the operation of more efficient social learning, due to imitation, and facilitate the discovery of more effective means of administration. Moreover, stringency in culture synergizes with the acquisition of a stronger cultural identity, which can dictate the values according to which a good life is lived, and consequently of hierarchies that impose them. In the feudalistic agricultural societies of Europe in the Middle Ages, the role of the Church in maintaining stringent cultural norms was prominent and played its role in a system that lasted for several centuries; during this period the population went from about 15 million in 600 CE to 100 million in the 14th century.

At the same time, generalized warfare became more common. This is attributed to two major factors. First, the resources that could be leveraged by the winning group in a conflict were more significant now, considering the magnitude of agricultural productivity, that justified the risk taken by the aggressors. This applied to various combinations of defendants and aggressors, for example when the aggressors were nomads attacking agricultural settlements or when one agricultural settlement confronted another with the ultimate goal of assimilating the other and benefiting from its resources. Another example would be when a society has a surplus, that is, considerably more resources than needed, and needs more land for the increasing population; or when new markets are sought to absorb surplus production. This last example is especially applicable in cases where the war and the desire to dominate are not physical but rather financial, yet the mode of operation is similar to physical war. It requires a certain culture to have come to be receptive to what the dominant force has to offer and effectively facilitate the assimilation. In China in the early 19th century, when trade with the West became significantly unidirectional, with the West being the importer and China being the exporter, in the absence of such a culture the imbalance was resolved by engineering the widespread availability of opium to the Chinese and launching the Opium Wars of 1839 to 1860.

There is another factor, though, that has contributed to warfare. It generates its own dynamics and usually receives less attention. Agricultural societies offer a strong sense of shared identity among the group members, because they require higher cultural coherence to satisfy both administration and the spread of progress. As such, the range of both historical and aspired needs of the people becomes both more intense and wider at the same time. In such groups, confrontation is likely, not only because more will be gained which in turn justifies the risk for greater losses, but also because within the spectrum of the culture-associated historical needs, a whole new set of ideas and ideals has been imprinted which make the will to engage in war, despite its consequences, likelier.

Such mechanisms underscored the expansion of cultures after the agricultural revolution and their transformation into the various civilizations that, regardless of their diversity in organization and ideals, engaged more and more people over time until they collapsed due to conflict or disintegrated due to the difficulty of sustaining them. Despite the negative consequences of being part of a larger group, especially if one was not a member of the ruling group (usually by birth-right only), large-group membership had its benefits. The notion that the oppressed were maintained as such only by force is in most of the cases valid for a transitional period only, and in the initial stages; but eventually a new *status quo* was generated that was in general acceptable until the conditions changed.

Naturally, in such complex systems of social organization, niches of distinct identity—national, for example—started to exist which satisfied psychological needs for one's individuality within a large-scale structure. The feudal organization of Europe in the Middle Ages, with its immobility and expansion of agricultural societies, resulted in the creation of national identities that, almost a millennium later, concluded with the creation of the contemporary nation-states. Progressively, as culture becomes more uniform, borders start acquiring a different significance; instead of marking cultural differences with more distinct ideals and mindsets, they convert into different trade zones among which similarities surpass differences.

As we approach the modern period, we can readily recognize this tendency to desire to live in larger societies, as evidenced by the globalization we currently witness. The internet in combination with the opportunities for online commerce have made one's physical location a rather relative notion. The "work at home" options that flourished during the COVID-19 pandemic also brought forward the "virtual office," highlighting the relative significance of physical location. People can now interact, imitate, and serve as models to others in different parts of the world, and as such they become carriers of cultures that converge. Cultural identity remains a live notion, but mostly it expresses the need for individuality rather than reflecting tangible differences in life perceptions.

A notable exception to this expansionist perception of groups, in historical times, is provided by ancient Greece, where people, for several hundred years until the integration of all Greeks into the Macedonian empire of Philip, the father of Alexander the Great, consciously selected to live in small-scale societies, administered independently, in the city-states they had established. They traded with each other and frequently engaged in localized warfare, but the idea of conquering adjacent city-states to create larger kingdoms was not a prevailing notion. The strategies pursued to cope with population growth and to secure financial resources included trade (primarily followed by the Athenians), or the establishment of colonies, pursued by the Corinthians. The Spartans, on the other hand, invaded adjacent cities—not to incorporate them into their Spartan city-state but rather to enslave their population, the helots, and acquire their resources. In these cases a coherent Hellenic culture persisted, and despite their distinct administration and perceived identity people could travel, work, and live in a different city-state from the one they were born in, being recognized there and playing important social roles. Probably, the strong cultural and intellectual traditions that had been established in ancient Greece managed to dissociate the financial benefits of group expansion from the cultural benefits of maintaining their distinct identities in their city-states, so that wars of expansion were not seen as a desirable choice. In that case, it seems that the different historical needs that were already established guided the dynamics of Antiquity, placing other priorities as the collective aspiration.

With the Romans, however, less than a hundred years later and despite their proximity to Greece, different ideals prevailed in their collective culture, and they engaged in systematic expansion that within 300 years led to the creation of the Roman empire.

Globalized humanism: All people matter at the same level, more or less!...

Being part of Western Civilization, it is hard for us to identify certain features that we consider more important than others. All its characteristics contribute to what we understand as the Western line of thought, and we feel that each of its components is indispensable in the formation of our collective identity. One characteristic that stands out, though, especially when we evaluate the social structures of the West and its aspirations for what we perceive as ideal forms of society, is its humanism. This humanism reflects the perception that people are at the epicenter, and that all of them should be recipients of progress. This must be made feasible on the basis of equality and nondiscrimination, both of which represent central elements of our current cultural identity. We maintain a deep conviction that progress, in medicine for example, should be available to all. The same applies to technology and to all the developments that make our life easier. At the level of social relationships or the relation of each person to authority and the law, the underlying concept, again, is that of equality, and that status does not translate into different treatment from the authorities and the laws. If we realize that this is not happening, either intentionally or unintentionally, we consider it to be a distortion of the system and we try to fix it. All progress is now seen through this lens, to eliminate inequalities that are associated with our birth status. As such we have carved into our collective mindset a deep conviction about equality, the belief that all people, irrespective of social status and income, should have unrestricted access to what progress has to offer. This is a direct consequence of the notion that "all men are created equal," which warrants equal opportunities for leveraging the outcomes of progress.

This is a direct consequence of the humanism of the Enlightenment, which aimed to put individuals, and not humanity as an abstract concept, at the center. This is not, of course, a discovery of the West. Prehistoric people enjoyed a version of humanism through the equality of the members of the group to which they belonged. Since the group mattered, all those who were part of it should be protected, defended, taken care of, and fed. In ancient China, despite its tight hierarchical structure, Confucianism also advocated humanism on the basis that each individual has a role in society, and therefore he and his works should be respected and valued. A functional society, according to Confucius is one that can integrate everyone and deliver a valued product. What was distinct, however, in the West, especially from the Enlightenment onwards, is the active effort to rationalize these ideas and enrich them with a firm and solid philosophical background that would eventually emerge as the robust imaginary that guides our future.

The integration of these notions into the collective mindset of the people is also reflected in the various utopias proposed during the Enlightenment, which imagined societies in which all people would enjoy privileges irrespective of hereditary rights, social status, or wealth. With the exception of the ancient Greek Polis, which also recognized the equality of its citizens, this was an unheard-of notion before these times, when the privileges of the higher class were not questioned.

Although these Utopias had not materialized, at least in a manner that would resemble the anticipations of those who originally proposed them, the sociopolitical dynamics they generated were sufficient to transform the political system and the ways societies are organized toward ways that promoted humanism. More importantly, they established this ideal as a guiding force for the future, in which the emancipation of people from any forms of oppression was central. Such oppression, depending on the period, might be racial, financial, physical, ideological, or more recently even cultural. Nonetheless, the relief of any plausible burden that is imposed remains our over-arching goal.

At the same time, a synergy was conceived at this point between humanism and reason. The latter, by offering the foundation for scientific and technological progress, was seen as the power that would enable people to attain liberty, by relieving them from what so far was oppressing them. Through reason we developed science and technology that relieve us of the oppression of limiting resources or environmental constrains, and through reason we address the oppression that is imposed to us by cultural norms or any form of ideology. As we will discuss later, though, these notions are somewhat incompatible because the disproportional advancement of science and technology compromises humanism and individuality. Science and technology generate complexity and promote another form of superstition that is based on trusting experts for the truth. In this sense they block people from seeking to pursue their own understanding of life, because of the overwhelming amount of specialized information that accumulates. Thus, scientific and technological progress rather compromise humanism instead of promoting it. Nonetheless, as ideals, both humanism and reason maintain their unique position in our collective imaginary in the West.

6. Menopause, the Elderly, and the Culture of Prestige

Menopause and the beginning of culture

All cultural developments that culminated in the ideals of the present society can be traced back to the origins of our cultural evolution. While the practice of imitation was essential for our cultural evolution, the development of models was a mandate that had to be fulfilled, to leverage this ability of humans to learn by imitating others. Menopause played an important role in this process because it assisted greatly in formalizing learning as a process that is integral for the society's function. It also contributed to rendering prestige, as opposed to dominance, as an avenue for the attainment of status.

To some degree, all primates exhibit the characteristics that we recognize as the prominent features of our species. Although we are fascinated by discovering these skills in our close phylogenetic relatives, the truth of the matter is that they are amazingly enhanced in humans, while in the rest of the primates they exist only in a minimal form. Nonetheless, their existence in these species highlights the evolutionary origins of our behavior and the trajectory that led us here. As a species, humans evolved in ways that made us capable of developing social learning skills, sustaining them, and depending on them. Our body and physiology evolved to leverage these skills. In our neuroanatomy, our brains have increased in size compared to those of other apes. Our behavior reflects that our children, for a considerable time after birth, depend on us both for feeding and protection, and at the same time for learning and integrating information that is already available without the need to rediscover it from scratch. We are essentially trained for that when we reside with and depend on our parents since the earliest time of our evolution. Our prolonged coexistence with our parents allows ample time for teaching and learning. This is also reflected in the fact that we are the only species, except for some whales, in which reproductive age is significantly lower than biological age. We are able to reproduce for only a fraction of our lifetime, and this is particularly true and clear in females who undergo menopause and partially true for males as well.

Women stop menstruating at around the age of 50-55, which is long before they age and die at around 85 years (according to their life expectancy in the West). This implies that in humans, there may be another evolutionarily selected function that relates to our elderly that extends beyond offspring production. Such a function is attained after reproduction ceases; in other animals, it is absent or minimal. From an evolutionary perspective, an intriguing hypothesis today is that being physically active and capable after menopause allows females to care for their descendants, directly benefiting them and their offspring (Austad, 1994; Takahashi et al., 2016; Hawkes et al., 1994). This has been nicely termed the "grandmother hypothesis," according to which grandmothers, instead of taking care of their own children (whom they can no longer produce after menopause), start taking care of their daughters' children, thereby increasing the chances for their survival. This hypothesis has received widespread acceptance and traction because it aligns with kin selection, according to which grandmothers share genes with their granddaughters; thus, allocating their resources is a beneficial strategy. Nonetheless, skepticism has been expressed, and concerns regarding the breadth of its application in different historical human populations have been raised (Moorad & Walling, 2017). Analogous reservations have also been noted regarding a related hypothesis, the so-called "Mother effect," which posits that women stop reproducing at a certain age to ensure that their younger offspring receive adequate care and rearing (Lahdenperä et al., 2011). If they continued producing children when they were older and less capable, the later children would receive less attention and care; by default, the younger ones would eventually have to grow up in the absence of their mothers, who will have died.

Concerns originate not only from theoretical considerations regarding the population dynamics of the group, but also from historical records. Studies of ancient DNA derived from tombs have provided interesting hints regarding the structure of prehistoric families. Among the findings of these studies is the absence of adult daughters from the groups. Although young females that share DNA with the group are designated as "daughters, the presence of genetically unrelated females was recorded. Furthermore, older women of high status were present, as indicated by the presumed rituals in their burials and the tombs' structure and location (Mittnik et al., 2019; Fowler et al., 2022; Rivollat et al., 2022). In these examples, grandmothers could not take care of their daughters' children since the latter were not present. However, the older women could acquire high status, as indicated by how they were buried. This elevation in status must have been related to a potential function they had beyond being mothers or grandmothers. Their daughters were entering other groups where they reproduced, and their children were cared for by them only or by other genetically unrelated females. Thus, menopause could not have played a role in enabling the older women to take care of their daughters' children. Menopause, however, evolved and became fixed as a trait in humans by other mechanisms that could have corroborated the proposed grandmother and mother effects.

How could menopause have been integrated into the function of the older females of the group and be selected for? It may very well be related to a role that is wider than what is dictated by motherhood and grandmotherhood, extending beyond the tight group of individuals with whom a pool of shared genes exists. An intriguing possibility is that it relates to their role as carriers and disseminators of information and wisdom that have accumulated over the years. In that case, older females might function impartially towards not only their own kin but towards everyone in the group, which also includes individuals unrelated to them. There are several reasons, both conceptual and physiological, that can support this.

Humans, contrary to other primates, show a preference for mating with younger females (Buss & Schmitt, 2019). In fact, in chimpanzees, a stronger preference for older females exists (Muller et al., 2006). This human trait of preferring younger females as their mating partners could very well explain the loss of reproductive activity of older females and thus would be associated with the paternal investment and pair bonding that, although present in other primates, are more prominent in humans (Morton et al., 2013; Muller et al., 2006). Older women, then, may have started acquiring another role, as models for imitation, once their role as potential breeders became less prominent. This would have been encouraged by the fact that half of their immediate genetic relatives—namely, their daughters and grandchildren—might be absent from the group. It seems that the role of models and educators, unique to humans, starts with older women and becomes more intense over their lifetime. This is an evolutionary adaptation that benefits the group, raising the value of the teacher not only as one of the oldest occupations but also as the first, probably, that has been evolutionarily selected. In that case, being a teacher is a biological imperative and not only a cultural construct.

If we hypothesize that women could menstruate and reproduce for longer periods in their lifetime, we could easily postulate that such functions as mothers and caregivers would absorb considerable time from their efforts to teach the younger members of the group the skills necessary for their survival and development. This would come at a cost to other roles they may have within the whole group and occasionally it would be contradictory to them, as it would exhibit partiality instead of the demanded impartiality for the group's success. Unavoidably, older women would show favoritism in teaching children with whom they have relatedness, compromising their function towards the other children of the group. However, when their children and grandchildren are old enough and do not require their attention, the older women can now focus their efforts on the whole group impartially.

It is not only the inability to produce offspring that relates to this new role, but also the various physiological changes associated with those hormones that, before menopause, sustain reproductive capacity, and after menopause are dampened. Estradiol, which is the steroid hormone that, along with progesterone, regulates the estrous cycle, is associated with several behavioral responses such as the induction of courtship and mating cues, the promotion of aggressive and impulsive behavior, and increased irritability (Horstman et al., 2012; Laredo et al., 2014; Jennings & Lecea, 2020; Wharton et al., 2020). These behaviors antagonize the function of females as potential social models. Consider that an effective teacher must be patient, with reduced aggression and devoid of mood swings, all of which are behavioral responses related to estrogen function or to estrogen fluctuations during the estrous cycle, which diminish after menopause (Spiteri et al., 2010; Oliveira and Bakker, 2022). Thus menopause, in parallel to the transition from partial to impartial care, may have also marked an evolutionary transition from dominance to prestige for the acquisition of social status, since dominance is typically related to aggression, intimidation, and coercion, all of which decrease after the estrous cycle ceases. Typically, dominance represents a more archetypal way by which the enhancement of status is achieved. By promoting prestige, a whole new set of dynamics was triggered in our cultural evolution, formalizing the role of teacher as a model for imitation, which led human societies to appreciate and value perspectives that were different from before.

Relief from the menstruation taboo

Besides the biological burden of pregnancy itself and the eventual caring for their offspring, there is another factor that must also have contributed to the elevation in status of postmenopausal women. In different societies and cultures, across the globe, menstruation has been associated with various taboos (Durkheim, E., 1963; Laws 1990; Guterman et al., 2007). Such perceptions have deep historical and anthropological foundations and they still occur today in different places of the world, albeit outlawed (Thapa and Aro, 2021; Gottlieb 2020; Garg & Anand, 2018). Rather intriguingly, it has been proposed that menstruation, and the taboo associated with it, was pivotal for our cultural evolution as it instructed among others, the initiation of solidarity strategies among the females against male sexual harassment, that peaked during the bleeding days (Knight 1991). The blood ties of the warrior men were finding their analogy in blood ties between menstruating women. Such strategies relieved, albeit temporarily, the reproductively capable younger women from coercion, that was exerted by the males, through their dominance. Additional norms developed along with that. The success of hunting was central in this process at which the isolation during the menstrual cycle, that is synchronized with the lunar cycle, allows men to remain undistracted while they hunt, and signals rewarding for the males that upon their return to the group bearing successful kill, will find their women done with the bleeding (Law 1986; Knight 1991).

From that perspective, the taboos that had been combined with female solidarity during menstruation, were providing a starting point for the development of various norms that regulated interpersonal and group relationships. In these norms, reproductive activity intermingled with hunting practices and celestial signals, increasing the complexity of human relationships and of their societies, and instructing the development of rituals that possess a spiritualistic dimension. Gradually these norms became even more complex, engaged male family members as well, and guided a network of relationships in the prehistoric societies.

The bleeding during menstruation, combined with the inability of women to perform marital duties at this period, has eventually resulted in treating women as impure when menstruating. Commonly, women were not allowed to participate in various social functions, they were prohibited from touching objects that were considered sacred and were occasionally isolated from their communities and from interacting with others, until their period ended. Characteristic for this attitude are the menstruation huts, at which women menstruate together in isolation, that still exist today albeit being outlawed (Adhikari, 2020).

During this period of their lives, that lasted for about 35 years from puberty to menopause, women were enslaved by their estrous cycle, which dictated their biological and social functions as prospective mothers,

mothers, or individuals in isolation from their community regularly for a fraction of their pre-menopausal adult lifetime. Menopause, therefore, alleviated the burden imposed on women by the taboos of menstruation and allowed them to participate in social functions uninterruptedly—functions that otherwise could not have occurred. Following menopause, they became liberated from the negative treatment they had to endure for a few days each month throughout their reproductively active lifespan. Now, their function and role in the group are not dependent on the time period, which is no longer a variable, and they can exert their influence over time without considering their monthly bleeding as a factor. This perspective is evident in various cultures, such as among native American Indians and others, where menopause is viewed positively. In these instances, menopause signifies the alleviation of the restrictions associated with menstruation, promotes inclusiveness in social functions, and has been linked to the acquisition of wisdom and the potential for an elevation in social status (Flint 1975; Madden et al., 2010).

Importantly, as a physiological process, the entrance to menopause is a distinct moment in time and is not progressive. This contributes to its treatment as a defined biological moment, marking a transition that signifies two different and distinct stages in women's adult lives: The phase during which their primary purpose is childbirth and care of children, and the phase in which they are no longer capable of parenting but can now serve the group from a different standpoint. They can operate as social models and help rear other group members, irrespective of genetic relatedness. The sharp distinction between these two phases, characterized by a discrete biological conversion instead of a gradual change, has additional value in the collective social psyche that challenges the notion of aging-related deterioration. Within a year or two during menopause, the same woman, who maintains the same physical capabilities with only her reproductive capacity being excluded, acquires a new social role and status. This transition is not signified by a progressive change linked to aging, but rather as an abrupt acquisition of different biological features and social roles.

The males and the elderly

Similar considerations also apply to men, whose reproductive ability does not cease but certainly decreases with age. Testosterone, the main hormone in males that stimulates reproductive activity, decreases with age. Testosterone is also associated with aggression and physical performance, both of which relate to dominance. Thus, as testosterone levels decrease over time, this corroborates the enhancement of prestige, which remains the only viable strategy for males to retain their social status. There is an additional consideration: If we hypothesize that males retained their full reproductive capacity at its full scale in advanced age, then, the absence of enough females in the group capable of reproducing would lead to strong competition among the males of the group, with disastrous consequences. Alternatively, if we hypothesize that males also undergo menopause, like females, then the preference for younger females as mating partners that triggers menopause, which in turn promotes the roles of older females in supporting their grandchildren and other group members, could not have been sustained. In that case, parental investment and pair bonding may not be prominent enough to support learning during child-rearing, necessitating quite different evolutionary strategies for our species. Thus, a balance is attained that sustains these dynamics, according to which females cease to reproduce around the age of 50, while males can still reproduce but with lower efficiency. After that age androgens, which are also linked to aggression, decrease in levels, thereby influencing human behavior towards a personality that is more consistent with that of a teacher-model.

A reflection of this is the prominent social role the elderly have in all societies across different cultures, which apparently is not simply a cultural construct devoid of biological significance. Rather, it is an evolutionary adaptation that benefits the group directly. In other words, given the evolutionary trajectory that *H. sapiens* has followed, it could not have been done otherwise; the development of social norms recognizing the role of teachers in the society's well-being becomes a necessity, operating as a safety valve that ensures the transmission of information will be valued and sustained. The prominent role of females as benefactors of the whole group is also evident in the various matriarchal legends of our past, where goddesses were deemed divine not only because of their ability to give birth but also because of their role in society that supersedes their own genetically dictated interests. Finally, in both the East and the West, great teachers and sages like Confucius, Laozi, or Socrates are always depicted as old individuals, aligning with the perception that they, at an advanced age, are the most appropriate to function as models for imitation. In contrast, Religious figures such as Jesus and Buddha were younger, probably to emphasize that their divine character was implemented early in their lives.

Another interesting observation that supports these notions comes from the comparison of hormonal levels in populations of different cultures. Eastern cultures are generally considered to be more hierarchical than Western cultures, with the elderly enjoying a special position of respect and prestige in society. Comparisons of sex hormone levels in Asian and Western descendants indicate that the levels are somewhat lower in the former (Xu et al., 2014a; Xu et al., 2014b; Wang et al., 1991; Schoenaker et al., 2014; Sowers et al., 2004). Although socio-environmental and other factors can influence this observation and it should be considered with caution, it is also indicative of the notion that the effects of these hormones are not compatible with their role as teacher-models. Thus, societies where these hormones are at lower levels in the population are more prone to developing hierarchical structures.

Altruism: Does it really exist?

In biology, altruism is defined as behavior that benefits others at one's own cost and is explained by kin selection or providing resources to genetic relatives, thus increasing their chances of survival. As soon as people began creating organized social structures, this behavior also came at a cost; however, the benefit is not necessarily offered to genetic relatives. Individuals delegate resources to others with whom they do not have an apparent genetic relationship, acting in a manner where a portion of their effort directly benefits others, while they only indirectly receive benefits. These benefits could be realized through offspring, genetic relatives, or the group, regardless of its size and scope. The different norms that guide social organization and are imposed by culture determine what portion of this effort will indirectly return to individuals and in what form. It could be in the form of genetic perseverance, material goods or services, or certain benefits that fall within the domain of historical needs. A well-functioning society should maintain this return in order to remain stable and prevent abrupt changes due to intrinsic or extrinsic factors. Essentially, it should reassure its members that the rules of the game will not suddenly change. When people become suspicious of society, believing that the expected return will not materialize, they become unwilling to invest their effort in a common good (Reich, 2018). Therefore, the appropriation of a portion of our efforts and resources to others, which highlights altruism, should involve a reciprocity where some benefit will return.

From that perspective, altruism becomes a direct measure of how broadly we define the group and its interests, and how we are integrated into the process, or, conversely, where "own" or "individuality" ends. A mother, for example, exhibits altruism toward her children when she carries and feeds them, as this reassures that her genes, through their children, will receive the benefit and persist. The same applies to a grandmother who stops reproducing to rear her grandchildren. This concept extends beyond that; when her elevated status through prestige prioritizes the group's benefits, the individual's altruism applies to the group and transcends definitions dictated by genetic relevance. The act of altruism, however, is present and understandable when we accept that the benefactor prioritizes their identity as a group member over their identity as a specific individual with a unique genetic signature.

When the group is defined in cultural terms, altruism takes on a new meaning because the beneficiary is everybody that is part of the same cultural entity. Thus, the success of the group transforms into the success of the culture, which acquires a stand-alone value beyond its people and sets the framework within which humans, after we are born, live our lives and establish our collective and individual goals. The culture acquires a value that is equivalent to our genetic heritage and may also contrast it. Today, as we will discuss in more detail in the next chapter, we have reached a point where our primary identity is defined by our identity as a human species within the context of our Western heritage. Thus, our altruism is related to the success of our Western culture and its spread to all people in the world. The abolishment of the priority of genetic relatedness as a driver for altruism was seen in prehistoric humans, before civilization emerged, even as early as the Neanderthals. Therefore, altruism is not purely a construct of different cultures or of the Western tradition, by which it has acquired a wider perspective, but rather a characteristic of our human identity that prioritized the evolution of culture over the benefit of genes and of the people who carry them. It is not that culture caused the development of altruism, but rather that altruism was already ingrained in our human identity as soon as our species evolved and was leveraged to promote the evolution of culture. Depending on historical needs, altruism has acquired different dimensions, covering the group that has expanded in size over time, from a band of genetic relatives to the whole species.

The unit of evolution

Ultimately, all questions involving evolution are reduced to what it is that evolves. Is it the gene, the individual organism, the breeding pair, since they require each other to reproduce, or the population? Is it the species, or the culture as an entity, that surpasses the biological substance of species and functions as a unit in its own right? The trivial answer is that everything evolves, so there is no simple answer to that. Yet, in a process that moves several entities, there may be one entity that is the primal one and precedes the movement of the others.

In contemporary biology, the gene-centric view of evolution seems to acquire more traction, as it appears able to consolidate different standpoints and unify the controversies that emerge when the unit of evolution is described. This view became highly popular after its introduction by Dawkins (1976) in his seminal book *The Selfish Gene*, and it puts forward the idea that it is actually the genes that are subjected to selection, while the cells and, eventually, the organism are just the vehicles that also change to accommodate the continuously emerging needs. What counts through the evolutionary processes is the net sum of their impact on the genes that determine what is beneficial and what is not.

Frequently in biology, to describe altruism, we focus on the competition that occurs between entities that belong to the same unit of organization. Cells frequently sacrifice themselves to benefit other cells of the same body or the organism. For example, the cells of the immune system have as their only mission to die in the process of eliminating threats that can compromise the other cells of the body, so those other cells can survive. All cells of the body actually have a definite lifespan and will not leave descendants themselves, but the cells of the germline will. If the mortal somatic cells do what they have to do effectively, the germ cells will have better chances to leave more and more efficient descendants. These descendants will carry identical copies of the genes that are carried by the mortal cells, so the mortal cells, through their sacrifice, secure the survival of their gene copies. The unit that receives a tangible benefit in that case is the organism of which some cells may survive and others may not, but the organism certainly does. The same applies to communities in the animal kingdom, such as in bees, where only the queen will leave descendants while the vast majority of the bees in the same nest will not. But again, by functioning efficiently, the hive will do better, and the genes that are shared will survive. In people, we can also recognize similar strategies, where certain individuals commit themselves to professions prone to sacrifice, while others abandon their reproductive privileges. Overall, there is an extremely high disproportionality between effort and the resources that are distributed among individuals, reflecting the inequality in our organization. What benefits, in the end, is the population and the society, depending on the perspective, that will do better in the future. In these cases, however, we can recognize that the higher order acquires a substance that exceeds the substance of its units, and its priorities and fate are different. The individual organism consists of several trillion cells that, nonetheless, except for our neurons, which are in the range of only billions, are not the same throughout the organism's lifetime. However, we perceive the organism as a unit, despite the fact that it constantly changes consistency. The population does persist, but its units, the individual organisms, perish.

The tendency we can see emerging at the evolutionary scale is that when complexity increases, and different units form structures which themselves also form structures of higher order, the beneficiary is always the structure of higher order while that of the lower order, the subunit, receives the greater cost. At the same time, as we move higher in this organizational scheme, the higher unit acquires an essence that is tangible, but it exceeds that of its subunits. A portion of the population of the lower subunits will receive benefits, but another portion will sacrifice themselves, and the net benefit will be received by the higher unit of organization, and through that, it will be passed to the lower-level subunits. This is consistent with an expansionist perspective through which we can identify a trend towards higher orders of organization, in which the costs in the individual units are offset by the benefits received by the structure of higher order. Thus, from within a benefit-driven perspective, higher organization will be favored continuously, and the tendency to enlarge the group is sensible. The fact that we, as unit-organisms, have to function and operate within the limits of our individual consciousness fuels the anthropocentric perception we have developed, through which we try to see and to explain everything in reference to our own individual dimension. We do not perceive ourselves as the sum of the approximately 30 trillion cells of our body; rather, we perceive ourselves as a single entity. While we also recognize our interdependency with the population we live in, we don't perceive ourselves as units of this population; rather, we see ourselves as individuals within the population. However, when culture is imposed, we start acquiring an additional identity that is reflected in our membership in the group and our historical needs but exceeds our participation in the population. This is because our culture and traditions possess a lifetime that exceeds us and our population, and they operate at a different sphere from the one we can appreciate as biological entities.

Following cultural engagement, quite safely we can posit that the unit of selection is the group, with whatever this notion entails. If we consider that groups tend to become larger over time, from that perspective evolution is intertwined with the expansionist perception. Culture acquires an essence that is self-sustaining and self-fulfilling, impacting individuals and populations but aiming at its own perseverance. The cultures that are considered successful over time tend to expand by engaging more people and populations. This is not done only through dominance, as it invades and absorbs other cultures, but also through prestige, as well; others try to imitate it. Prestige is mostly responsible for the imprint that a culture leaves in history and can provide a foundation for delivering other cultures following it. The formidable Mongolian empire at the beginning of the first millennium, which derived its power for a few hundred years from its dominance, had very limited, at least direct, impact on subsequent cultures. In contrast, the City of Athens and the associated Greek city-state, while having only a very limited impact through dominance and a population of a few tens of thousands, had a tremendously high impact in terms of prestige that is admired and remains influential today. Thus, where a culture relies on prestige and the corresponding group of people becomes larger and larger, it is considered successful. As we will discuss in a subsequent chapter, this is precisely the basis of the fact that today, we tend to live in a single culture, and a group that is defined by our shared identity as a species. It does not really matter where one is from or where he or she lives; they are really members of a single group of people that, irrespective of geographical location, are justified recipients of the progress that will resolve their fundamental biological concerns on a more or less equal basis. Even if this is not the case today, the trend is that it will be the case soon.

From that perspective, we can recognize a progressive change in the unit of evolution that is consistent with its continuous expansion. Initially, in the earlier biological times, the unit of evolution was just the gene, which established partnerships with other genes for their mutual benefit. As biological organization evolved, complexity increased, and eventually, humans appeared, along with their cultures. While initially, evolution favored groups of humans to adapt better in their specific environments, the mandates of our culture, especially during the last few hundred years, have resulted in the whole gene pool of *H. sapiens* being targeted by evolution. There is a difference, though: Instead of being the unit or the subject, it is now the actor of evolution. Our efforts today, instructed by our culture and enabled by our technologies and our science, within the ever-expanding groups we live in, aim to maintain the whole gene pool that is available and prevent its change. We wish to be able to allow every gene to persist in time, irrespective of its actual impact on individuals. Culture dictates that every individual should benefit from progress, and technology and science enable this. In the societies we live in, we try to ensure that biological success becomes irrelevant and dissociated from our reproductive efficiency, and the cost in fitness at specific genes is offset by the benefits that science and technology provide. Culture, in that case, is brought forward as the major determinant of the new dynamics that have emerged because it establishes a structure that exceeds beyond the genes, individuals, and their populations. It is not external, though, to the people like the environment that requires people to adapt to it. Culture is inherent to the people, but instead of selecting for, like the environment does, it selects against potential changes. Although norms and expressions of human activity change, the main ideas and ideals of our current culture have a resilience that discourages change. Culture now has acquired a dominant role over biology and raises again the question of how different it is from biology since it is inherently associated with it. The group, through its cultural dimension, instead of being the subject of evolutionary changes, dictates evolutionary stability by enforcing genetic stability while culture itself advances.

Interestingly, this effort does not apply only to the human gene pool but also to the gene pool of all organisms that live on the planet, through our systematic efforts to conserve biodiversity. It is rationalized by a combination of our guilt for our impact on the environment and our desire to bequeath our children a world that is at least not worse than ours. Irrespective of our motives, it reflects very well our perception that the group we belong to acquires a larger and larger scale.

7. Social welfare and care for the underprivileged

Welfare becomes wider

People have taken care of their sick since prehistoric times. The prehistoric findings of individuals with disabilities who were buried together with other community members testify to this. In ancient times, we start seeing the establishment of specialized institutions that were charged with the mission of providing assistance and care to those in need. Traces of hospital-like structures can be identified as early as in Mesopotamia at the end of the 2nd Millennium, as well as in Asia, supported by the spread of Buddhism. In most of these cases, care was linked to metaphysical and spiritual practices, in which the divine was involved in patients' care. In the ancient Greek world, we see the development of rational approaches—the basis of Western medicine—in patients' care. Hospitals, in a more modern sense, first appeared in the Roman world and were supported by Christians. Those originally took care of messengers and pilgrims and later evolved into structures that treated sick people (Cilliers and Retief, 2002). In the Islamic world, the first general hospital was built by the caliph Harun al-Rashid in Baghdad in AD 805 (Nagamia, 2001). Hospital-like institutions that were supported by state funds and policies were also established in China at the end of the first millennium by the Northern Song (Goldschmidt, 2023).

Care for the sick and elderly became a central component of state policies in the West in the late 19th and at the beginning of the 20th century, and this was strengthened by the Enlightenment-promoted humanism, at the same time that capitalism was being shaped. Capitalism, by demoralizing the accumulation of wealth, promoted inequalities that fueled the conflict that exists between liberty and equality, and the fact that each one grows unavoidably at the expense of the other. This mandated the implementation of special provisions for individuals in higher need and for those who were not part of the active workforce, such as the sick and the ones that stopped working because of their age. Humanistic ideals prompted that in organized societies all group members should be taken care of, at least as regards their basic biological needs for preservation, even when they cannot contribute to the group directly and be part of the active workforce. Of course, this attitude also had another dimension, that of reducing the social tension and providing a sense of security to those who may eventually require support during later stages of their lives. Nonetheless, it was founded on the ideal that everyone in the group needs to be taken care of.

The Prussian Chancellor Otto von Bismarck, in 1883, is credited with establishing the first healthcare system in the modern world, conceived according to contemporary standards (Butticè, 2019). Workers, through deductions from their wages, and their employers, through their contributions, sustained the so-called "sick funds" that covered expenses for hospitalization and treatment. Since then, similar models have been applied to provide healthcare to the population. In modern Brazil, healthcare is free for everyone according to the country's constitution, which views healthcare as a universal right and the government's responsibility. Different countries worldwide, using various models of state, employer, and employee contributions, render healthcare accessible to everyone. Models used in Scandinavian countries, as well as in Asian countries such as Singapore, Japan, and South Korea, provide high-quality healthcare today by applying free healthcare policies. By 2030, all Member States of the United Nations have agreed on the 2015 Sustainable Development Goals to achieve Universal Health Coverage (Tulchinsky, 2018). From that perspective, healthcare for everybody—not only for those in need who will directly benefit from such a policy—emerges as a desired need for our Western society.

Besides healthcare, Bismarck can also be credited with implementing the first pension system in the modern world. In a process monitored and supervised by the state, workers contributed to a fund from their wages that would provide a pension after they retired at the age of 70 (Scheubel 2013).

Liberty and Equality

Equal access to material goods and services contrasts with the liberty to pursue, through work, increased access to these material goods and services.

Since such goods and services are available in definite amounts in a given society, their disproportionate distribution will eventually compromise the principle of equality, making them unavailable to others. This becomes even more striking when the benefits of liberty apply to those we care about, like family members and dependents. Eventually, the dependents of successful parents will receive benefits through their parents' efforts that are not justified by their own actions. The liberty enjoyed by one generation contrasts with the equality applied to the next. This creates conflicts that demand resolution. Western ideals trap individuals in a Procrustean state in which they try to pursue two distinct ideals, each of which compromises the other: We want to be free to pursue our individual goals, but we also want all members of our group in society to receive benefits, at least at a minimal, essential level. In our society, it is acceptable for people to practice their own liberty and receive justifiable benefits, but it becomes more controversial when such benefits are applied not to the same person but to his or her dependents. In that case, liberty is restricted.

Take the public vs. private school system as an example. It is deemed acceptable for financially successful parents to enjoy the privileges of their wealth for their own benefit. However, what about when their goal is to use this wealth to provide better education for their children, thus making them more competitive and increasing their chances of becoming successful as well? Besides limiting our liberty, this stance has another consequence that interferes with the established trajectory: It restricts our altruism towards our kin, while simultaneously redirecting our resources and efforts towards other members of the group impartially. This is reminiscent of the effects of menopause we discussed earlier, where older women, free from the burden of caring for their kin, can care for other group members without a genetic relationship. It seems that our society, at this stage, becomes postmenopausal in its trajectory, which is also reflected in its other facets. It discourages radical changes by absorbing different positions within the operation of liberal democracies and actively engages in educating its members towards these ideals through prestige and role modeling, thus expanding the size of the group. At the biological level, it promotes the perseverance of the whole gene pool by dissociating biological efficiency in function from genetic heritage, thus preventing evolutionary change via selection (Kiaris, 2022).

The trivial resolution of such disputes between liberty and equality is that equality should shift from equality of access to equality of opportunity. People should be free to pursue their own goals, but society should ensure that everyone has the freedom to pursue their goals as well, irrespective of their background. By comparing American and European capitalism, we can identify this difference: the former favors liberty while the latter favors equality. The balance between the two changes over time and place, but its shifts provide systemic stability and maintain consistency with the overarching Western ideals.

The quest for equality does not mean that we live, or aspire to live, in societies that are devoid of any forms of inequality. We recognize that inequalities are intrinsic to our society, especially within the context of our liberal, capitalistic democracies, in which the accumulation of wealth, and what is dependent on that wealth, has been demoralized. Beyond this demoralization, which fuels abundance in material goods, the pursuit of wealth provides an additional benefit: It contributes cohesiveness to our societies since it offers a widely accepted measure for everything and a robust system of values that are quantifiable and objective. Work, suffering, and achievement find their equivalency in wealth. If one performs well, he or she will be compensated with a financial bonus or higher income. Also, if someone suffers due to the actions of someone else, he or she will also receive financial compensation. Our collective work effort has its equivalency in compensation, and promotion in status relates to the acquisition of higher income. Thus, financial compensation becomes central in the society. Unavoidably however, this fuels inequality that is now morally justified and accepted. Nevertheless, we view these inequalities, as long as they emerge within the context of well-organized and lawful societies, as the drivers of our economy and progress, ultimately benefiting all people by maximizing collective wealth. At least this provides a resolution to the intrinsic conflict that, on the one hand, we demand equality, but on the other, we compensate people differentially according to their effort, achievement, or suffering.

Along these lines, the moral foundation of our capitalistic Western societies is based. Progress benefits some groups more than others; nevertheless, all are benefited to some extent. Thus, progressively, a series of privileges advances from the status of desired needs to that of historical needs, and as such, they should become accessible to an increasing number of people.

More people access essential needs

Progress is reflected in this advancement of needs from the status of desired to that of historical needs which benefit more people over time. These include the basic biological needs such as food and shelter, a series of technological advances that improve people's well-being and compensate for weaknesses posed by biology and the environment, and access to health care. In addition, they include privileges regarding our participation in decision-making, social mobility, and education, all of which are interconnected. For example, it is widely accepted that medical care should be available to all people. No one will argue that only some people should have access to the best available drugs or visit a hospital when they are in need. It is just a matter of distribution: who will be able to do so, given the limitations in hospitals, physicians, and drugs available. Unrestricted access is not the case yet, since one has to have insurance and thus a job to access healthcare, especially in the American version of capitalism. And even if you do have access to it, this does not mean that you will enjoy the best health care possible. In any case, we recognize this problem, and we are trying to fix it and establish a system in which, eventually, at least life-threatening conditions will be dealt with irrespective of one's income and social status. In European countries, especially the wealthier ones in Northern Europe and Scandinavia, these goals have been approached more than in the rest of the Western World, but the trajectory has been established for all, in the West and the countries that follow the Western model. The "cost," of course, is that in Europe the difference between rich and poor is not as striking as it is in the U.S., and that the super-rich in Europe cannot exist in the numbers that they do in the U.S. Thus, while liberty is favored in the U.S., equality is prioritized in Europe. Nevertheless, the trend and the widely accepted social demand, is to develop societies in which care for the sick will be widespread and impartially the essential needs will be provided.

All this comes with a cost. The main instrument the state uses to support those in need is to introduce programs such as Medicare and Social Security. In the U.S., projections suggest that funds will be available for Social Security to compensate eligible beneficiaries until 2037. After that, unless there is a significant influx of funds, pensions will be restricted. Analogous and even worse is the case in other Western countries, where lifestyle changes have created a drive for earlier retirement, thus reducing contributions to the common fund from which pensions are paid. In addition, the working population is being reduced because of the increasing automation and the overall decrease in childbirth. As the expected lifespan increases, retirees and non-working people will also increase. Immigration may appear to be a solution for these Western countries; however, it results in unavoidable harmful consequences in the countries from which the immigrants come. In order to immigrate and be accepted in the new country, one has to be competitive enough to assimilate and integrate into the workforce. The loss of such competitive workers causes a so-called brain drain in the countries that lose them, increasing inequalities and reducing the prospect for growth in those countries.

Despite these issues, which can be viewed at least from an optimistic perspective as a distortion of the current state of things and thus resolvable, we can easily envisage a society in the future, in the context of a liberal democracy, that will function the way we anticipated, and that if you get sick you will receive treatment. The same applies to education, which in the U.S. represents a major point of concern considering the outrageous cost of attendance at both private and public Institutions. Nonetheless, since strong criticism and social discontent exist, it is also plausible to argue that sooner or later, provisions will be made to eliminate or reduce this main obstacle to social mobility as well.

These are the main ideals toward which the Western world is driven, and eventually the whole world will presumably adopt the Western ideals in the long run. One's income and social status will not be a burden. Whether this will be the result of a national health insurance plan, charities, or some unanticipated medical advances that make the treatments cheap and accessible to all is irrelevant. Regarding education, it might also involve forgiveness of state loan programs or tuition waivers supported by state or private sources. But It is the progress in technology that enables economic development and enables these advances, and that will make the rich become vastly richer—yet the poor will eventually become at least slightly better off as well; this, in absolute terms, will raise their health care privileges and access to education. The fact that this may not be happening as yet, at least to the desired degree, is considered to be a distortion of how the system operates, and it is therefore considered to be fixable.

Analogous considerations apply to the technological advances that make one's life easier and safer. Living in a house that is warm in winter and cool in the summer, protected from physical disasters and—even if this is not possible, provisions of the society will protect you—all this is considered a privilege. It is more and more common to hear citizens' complaints when a disaster strikes. The accusation that the "state was not there" implies that the state ought to be on hand and protect its citizens when a hurricane strikes or an earthquake hits. Probably, to some lesser extent this attitude is justified by the fact that those individuals that were affected have paid their taxes and thus renewed their group membership, by purchasing in advance their protection from the state like in a form of insurance. It is much deeper than that, however, as it relates to our collective ideal that the society, in the form of the state, should be present and able to support those in need. This reflects the conviction that the state, with its resources that were made possible through the technological progress and the collective humanistic culture we have developed, should take care of its citizens. And it remains a fact that this is the trend. Over time, more and more people have enjoyed the privileges of progress, which now extend to an increasingly larger segment of the population.

There is of course a long-standing debate about this optimistic approach and, on the fixability of the problems within the current system, generating criticism from various political perspectives. Inequalities among nations may fuel wars and conflicts, while internal inequalities can lead to class conflicts. However, this criticism does not call into question the validity of the ideals but rather the means of achieving them. For the purposes of this discussion, we may set aside this debate and choose the most optimistic outcome, assuming that systemic distortions will eventually be addressed, possibly even through the path we are currently following. It is also important to keep in mind that regardless of how lives may be organized, any changes will occur in a manner that does not affect the chances of individuals reproducing and passing their genes to the next generation.

By hypothesizing that the central ideals that emerged during the Enlightenment, which promoted humanism, will continue to prevail, we can make several assumptions about our social future. We can assume that even though the rich are becoming extraordinarily wealthier, the poor will also receive an increasing share of the benefits of progress. Ultimately, we can assume that, whether within the current system or in a yet-to-be-developed form of social organization, all people will receive a portion of the privileges that will, over time, improve their quality of life. Based on our deeply ingrained humanism, we find it immoral to live in societies where disparities exist regarding basic individual needs, and we are on a trajectory to address this.

In addition to medical care, progress is expected to eventually eliminate several other problems humanity faces, such as hunger and the ability to avert disasters and confront catastrophes. We have built strong confidence in the capabilities of our technology-oriented society, which is capable of solving all emerging problems. Our collective efforts are directed toward this goal. Ideally, these advancements will occur within a historical context free from major catastrophic events that could compromise people's well-being. Such events may include global physical disasters due to environmental factors, wars, and other crises. Whether these events are inherently related to the current trajectory is debatable. For example, it is argued that wars stem from global inequalities and the uneven distribution of wealth, linking them to our current mode of social organization, which may be unavoidable unless the entire system changes. Conversely, some argue that wars can be prevented, and their occurrence is merely a result of uneven societal advancement.

Similarly, environmental issues, despite being manageable through corrective and regulatory measures, are also associated with our wealth-seeking economies. Thus, they too are part of our current historical process and may ultimately lead to disasters, despite efforts to mitigate them. The alternative perspective suggests that these issues can be avoided if appropriate actions are taken early enough. We will address these concerns subsequently, as their impact could also influence our biological future. In summary, regardless of our material circumstances, reproductive success is ensured to the extent that every individual in the group can reproduce. Furthermore, reproductive success is entirely independent of social status. These factors will be the primary determinants of our species' biological future.

8. Globalization: one culture—one population

The term globalization was introduced at the beginning of the 20th century as a translation of the French term *mondialisation*. It became more common later on, and mainstream in the 1990s after the collapse of the Eastern bloc and the new state of world trade. It was perceived with both positive and negative sentiments, depending on the political angle the criticism was coming from. Some viewed it as an opportunity for expansion and growth, with new markets and production, and others saw it as a threat to cultural heritage and a decline in the quality of life due to intensified financial and social Darwinism. The consequences of an early version of globalization can be seen in the effects the opium trade had in China in the beginning of the 19th century. The West, led by the British, in order to offset the trade balance that was shifted towards China, implemented the trade of opium that devastated Chinese society, resulted in the Opium wars, and marked the collapse of a 3,000-year-old civilization within less than 100 years.

Despite the criticism, globalization reflects the new world order where boundaries become less rigid and the mobility of people, goods, and ideas, increases. Unavoidably, this impacts our perceptions regarding the group we live in. It is no longer just our country or the nexus of countries that we are in good relations with, that defines our identity, but practically the whole world, a fact that is also assisted by the discovery of new information technologies and the further development of transportation technologies. The world suddenly has become much smaller because now, one can live in one place and work virtually in another, sell the goods they produce to distant places, and use goods imported from elsewhere for their own lives.

More importantly, one can live in a culturally globalized environment, exchanging information instantly with people located in different places and appreciating the ease of communicating and connecting with others. The common foundation that is being developed, assisted by the new technologies of electronic communication, eliminates physical distance as a barrier. Consequently, the world is becoming increasingly homogenized, and the differences between various places regarding ways of life are becoming less strict.

In biological terms, globalization has also had a significant impact because it has resulted in the drastic increase in the size of the population we live in and within which our genes can be exchanged. Earlier on, intermarriage between individuals belonging to different cultures was more restricted, not only because of the cultural traditions and the taboos that may have prohibited it, but also because communication was limited. Now that the cultural foundations become more common, intermarriage between people of different traditions becomes more feasible. This change collapses significant barriers to gene exchange between populations and gene flow becomes less impeded. This new trend is expected to become even more prominent in the future and will likely result in a state in which any two individuals of different backgrounds will have higher chances than ever to reproduce, enabling new combinations of genes to materialize. Coming back to the impact of the emerging new cultural heritage, mate selection should ideally be driven by personal choice and not by any other social or cultural factors. At least this is the aspiration, to the extent that the society we are developing will advance to such a level that it discards taboos regarding marriage across different classes, traditions, and cultural heritages.

From the perspective of who is entitled to receive care and support within this new enlarged group, another interesting trend is also strengthened. People always, since very early in the history of our species, had the tendency to help those in need. We already discussed the fact that in all civilizations, a common trend—and at the same time, a manifestation of their success—was the care for the underprivileged. The extent of the needs that were covered was typically a function of historical time, reaching today a point that such benefit has become larger than ever. And there is also an additional factor that plays a role. Aligned with the humanistic ideals of the Enlightenment, everyone is understood to be entitled to this benefit, importantly those from different cultures as well as the dominant group. As long as someone is part of the huge group of the new globalized world, he is considered entitled to the benefits. Today, albeit it to a very tiny extent, the wealth of the West is extended to provide for people in need, to those that are typically classified as third world countries, in the form of humanitarian aid. Although this aid is too little to have a truly significant impact, it illustrates very clearly the perception that "the group" is much larger and that all people in need should be taken care of, irrespective of their culture and their traditions.

Importantly, such help is not provided only in the form of medicines, foods, or infrastructure to provide shelter and address the essential biological needs, but also in the form of instruments that target their culture as well. For example, they receive books and teachers, training opportunities for education on site or hosted in major Western countries, and even in the form of religion. Christian missions are active in these countries and besides their assistance in attending to the biological needs of the populations they visit, they also export their religion and their cultural traditions. Our perspective as Westerners towards other cultural traditions appears similar to the role that older women gain after menopause, which allows them to assist others without bias and to educate them so that they can ultimately become competitive within the community.

The actions we are taking significantly intensify the cultural effects of globalization and indicate that the trajectory we have established will soon lead to a more uniform culture, primarily based on Western traditions and ideals. The world is becoming culturally unified and Westernized. Therefore, future societies are likely to resemble those of the West rather than other cultural traditions. While the Western world also evolves and incorporates elements from other cultures, its core values and ideals remain distinctly Western, following the trajectory set after the Enlightenment. This trajectory supports the absorption and adaptation of various ideals for its own continuity.

These efforts receive a lot of criticism from different parties. On the one hand, they are seen as an intrusion and a threat to the local cultures, leading to their elimination. They are also seen as interventions that in the long run, instead of minimizing the problems, will expand them. From the perspective of the West, providing this support is a way to alleviate our guilty conscience for the problems we, as Westerners, have caused through our present and past policies. For some donors, whether individuals or corporations, it can be an efficient way to reduce tax payments while endowing oneself with an air of nobleness. This sense of nobleness and generosity is related to the savior's complex that has characterized the West in modern times. Nonetheless, irrespective of people's deeper motivations, these ideals are deeply engraved in the peoples' collective conscience and demonstrate that everybody in the world carrying the human identity is entitled to such help.

9. The Western tradition

Humankind has no nature, only history.
—José Ortega y Gasset (1883–1955)

When we discuss cultural evolution, up to the present day's globalization, this approach possesses a teleological dimension. It implies that we, the ones making this statement, acknowledge that a certain path has been taken which has reached a terminal point beyond which we cannot go further. Similarly, when we criticize the past and attempt to draw generalized conclusions, it carries the implicit notion that what is currently ongoing is of lesser significance than what has already happened in our past. If major, game-changing events are to occur in the future, what is the point of trying to assign significance to things that have already taken place? We may certainly assign some relative significance to past events, but this significance can be minimal if future events hold the potential to surpass it.

Besides this subjective opinion that justifies the teleological approach in our history, there is another reason that legitimizes our perspective in considering events of great significance that have already occurred: the sense that we have developed a system with mechanisms in place to prevent abrupt changes. Progress gradually moves towards a given direction, and we believe that the ideals we have developed will ultimately be satisfied in our future.

We assert that we have reached a point where sufficient wisdom has accumulated to justify retrospective criticism. This is a typical attitude we maintain as products of our culture, which legitimizes our criticism of everything we consider distant from us—geographically, chronologically, or culturally.

Whenever we evaluate the past and try to forecast the future, we always allow space for events of great significance that may arise and change the course of things. However, we do that mostly due to academic rigor rather than a genuine belief that such monumental events may indeed happen. At least regarding our cultural and philosophical path, we are confident that we have developed an ultimate system capable of integrating disturbances, fluctuations, and new developments in a manner that will not compromise the future trajectory we have established.

Although this bears a rather pessimistic perspective, it may not be far from the truth. Our whole intellectual stance at present has been developed in a manner according to which we are now bearers of a whole truth that enables us to see, in an unprecedentedly objective way, how the physical world functions and what the genuine impact of historical events was. The intriguing point from that perspective is that we are, in principle, receptive to the possibility of game-changing ideas that can drastically change how we think and understand the world. Yet, we completely block their occurrence and minimize their impact because to accept them, they have to be fully aligned with our established mode of thinking that we accept as valid. Therefore, such events cannot be truly game-changing events in their essence; instead, they will have to be components of the current state of things.

If we consider biological evolution as a tree, at the very top branch, we will have humans, close to the other great apes, while far from humans, we will see other species from different kingdoms and genera, organized in a manner where physical distance within the tree reflects evolutionary distance from the point at which common ancestors can be identified. Such a tree has been reproduced with remarkable accuracy by combining both paleontological, genetic, and anthropological evidence. The evolution of culture can also be depicted as a tree that branches out from humans, having perhaps some small roots and minute branches in other apes as well, but they haven't grown enough to be visible. In the case of the other apes, and maybe of the other animals, the distinctive feature is that information, although it can be transmitted, does not accumulate between generations. This is what differentiates the handling of information between us and our immediate evolutionary ancestors. Each primate, unlike humans, must at each time start almost from scratch in its journey through life. By studying how they live now, we can conclude with certainty how our evolutionary ancestors lived 1,000 or 10,000 years ago. The same applies to all other animals as well that start their lives having at their disposal only their genetic heritage and the given external environment in which they were born. However, studying how humans live today tells very little regarding how we lived, not 1,000, but even 100 years ago unless we have direct historical evidence for that or resort to isolated populations that, with assumed certainty, live like they did in the past. The identification of behaviors that persisted throughout the years in people, reflecting ancestral behaviors, demands a certain degree of speculation that is not always accepted.

People start their lives defined not only by their genetic heritage and external environment but also within a given cultural context that shapes them and is integrated into their lives. The Spanish philosopher José Ortega y Gasset (1883–1955) illustrated this by stating that "Humankind has no nature, only history," although he completely eliminated the impact of biology on people (Binder 2019). We now more accurately conceive the human condition as being constituted by three variables: the genetic makeup of individuals, which reflects their biological essence; the environment that modifies how people live their lives and with which they establish a reciprocal relationship, changing it while also being changed by it; and finally, their culture. The cultural component acquires autonomy in its existence and its own dynamics, as it determines how people live their lives, accumulating over time in an irreversible manner and gaining dominance over other constituents. The way we live now is primarily impacted by our culture rather than by our genes or our environment. As we will discuss towards the end, our existence is shaped synchronously by our culture and our biological essence, if the former does not in fact take precedence over the latter.

A culture of cultures

In the tree of cultural evolution originating from humans, the foundational cultures of our forager ancestors form a significant part near the base. From these roots, we can envision branches representing various major cultures and eventually the civilizations that have emerged over time. These major branches popped out after the agricultural revolution, about 10,000 years ago, especially during the historical period. At several points, these branches interlinked, similar to the evolution of species. Certain branches diverged and developed into distinct cultures that acquired distinct identities while still sharing common features with their predecessors.

Principally, though, what we now recognize as different civilizations—the majority of which are extinct—developed independently in various locations and times across the globe. Naturally they all stemmed from pre-existing cultures, but during their evolution each acquired a distinct identity, leading us to view them as separate entities. For instance, the practice of cooking food and the use of fire were not necessarily inventions that some populations created, forgot, and then re-invented. It is plausible that our ancestors consumed both cooked and uncooked food, but the consumption of cooked food was more efficient and safe, so the practice was widely adopted once the cooking process was discovered by different communities. Since this practice was associated with survival, it is likely that it was continuously practiced. Thus as people dispersed and developed their cultures, cooking became a deeply ingrained cultural element.

Different environments and resources influenced how cooking processes evolved, and these processes led to the establishment of distinct traditions. These traditions became so integral to cultures that they were often taken for granted. People were practicing them without being able to dissociate the act of cooking the food to make it safer and more nutritious, from the rituals that were associated with the whole process. Take as an example the spices used in cooking around the world, particularly by populations near the Equator. Historically, their use primarily served antimicrobial functions. In various environments, due to differences in spice availability and in the abundance of microbes, people started using different spices, and integrated them into their cuisine, fulfilling similar protective roles. Over time, these spices became associated with particular tastes that now seem intrinsic to different culinary traditions, such as those found in Indian or Latin American cuisines. People from these cultures have developed a preference for using spices like curry or chili peppers in their meals, often without consciously relating it to the origins of this tradition and to their original antimicrobial purpose. While people may be aware of these properties, their culinary choices are not driven by that knowledge. This practice continues even among immigrant populations in new environments where the original microbial threats are absent; they still rigorously use the same spices. This illustrates how information and traditions are passed down through generations, shaping specific cultures while remaining highly subjective.

In parallel, besides the specific processes that possess a causative role, at least as far as we are able to infer and appreciate them, a series of additional actions and processes were also developed that do not possess causative significance but fulfill an understandable role and are followed with the same rigor. These evolve into various rituals that offer a special character to different cultures.

Characteristic are the rituals that follow someone's death. The appreciation of the significance of death in someone's life plays a pivotal role in cultural evolution because it signifies the inception of an eternal soul that persists after life. From that perspective, people deal with the possibility that their existence supersedes their material substance as natural beings, and they appreciate that their biology is not the only constituent of their existence. Burial is usually preceded by mourning. Such mourning is distinct, intense, and follows a very precise protocol, and lasts for a few days. Occasionally, designated individuals, the mourners, are also engaged in guiding the process in addition to the priests who participate and orchestrate it, along with family members, relatives, and other members of their community. These mourners are not related to the family that experienced the loss, but they are experts in this role as they cry, sing specialized songs, and create the aura that characterizes this event.

The causality of this procedure is dual: First, it relieves the emotional pressure inflicted upon the survivors of the deceased. This process provides an acceptable outlet to relieve their sorrow in an environment that offers strong psychological support. In this way, the process illustrates the association between biological and emotional well-being. Second, it allows ambient time to pass to rule out the possibility of auto resuscitation or the Lazarus Syndrome, where, rarely, people can wake up after being declared dead (Adhiyaman et al., 2007). Both mourning as an outlet for sorrow and the delay of burial possess biological significance. At the same time, the whole process of mourning is tightly linked to a series of rituals, including special clothing, for which causality is hard to attribute. While, for example, in Western Christian religions and practices, black is the color of choice, in East Asia, all white clothing is preferred. One should expect the same color to persist in different cultures if it were associated with specific causal effects on the emotional state of those who mourn. Nonetheless, these rituals are practiced with high stringency and collectively constitute traditions that signify the respective cultures.

Rituals and traditions accompany significant life events such as death, marriage, birth, parenting, graduation, retirements, and others. Our lives are deeply intertwined with these traditions, which vary across cultures and serve to signify cultural identity. For example, when we meet someone and greet him, in the West we extend our hand for a handshake or even kiss (twice, in some cases), or we simply hug. In the East, however, bowing the head respectfully is the most appropriate way to greet someone. A rational explanation for the origins of these actions can be proposed. In the West, raising a hand signifies the absence of a weapon and thus conveys friendly intentions, whereas bowing in the hierarchical East indicates respect. However, these interpretations may be speculative, reflecting our desire to find causality where is none.

There are numerous examples of such rituals, as one immediately realizes as soon as he travels to a different country. Many of them can be understood as outcomes of significant cultural trends. For example, the act of bowing in the East is a direct consequence of a social structure that values hierarchy, where respect should be immediately extended and denoted, as opposed to the West, whereas hugging and kissing are used to promote a sense of equality and informality. In other cases, though, rituals just persist and continue to be followed, conferring coherence to the corresponding societies, as components of the cultural identity, fostering a sense of connection among those who share the same cultural background.

We can easily identify a noticeable trend today for the reduction of the impact of traditions on our lives. This is an outcome of the increasing power of rationality, which renders practices lacking a practical foundation increasingly meaningless. This relates to the power that technocracy has today, too, in that it makes science and technology seem like the only viable avenues we have for moving forward. It serves another purpose, as well. As traditions loosen, they become more accommodating to diverse cultural backgrounds, creating the perception that our group today is larger than what it ever was. This is not a matter of geographical proximity; for example, Australia is geographically closer to Myanmar, yet culturally, it is far closer to the United States. This has to do with the cultural identity of the people who colonized both countries. A prerequisite for the expansion of the group is to be more tolerant to diverse ideas, so more people can now be accommodated into them. The tolerance of both the Roman and the Ottoman Empires that they exhibited to their subjects are recognized as central elements for their success. We live this today in the U.S. in its extremity, at which tolerance of diverse ideas and the cultures they represent, have acquired fundamental significance in what is perceived as the emerging U.S. culture. The cost of this, nonetheless, is the loss of central cultural elements that may function as reference points and offers cohesiveness, and their substitution by a culture of neutrality against everything. Relevant to that is that the perception of being a group member becomes increasingly less potent and is substituted by a more superficial interaction, a union against common goals, if and whenever such goals can be identified, for our own individual benefit. Everything in this new emerging culture is becoming increasingly acceptable, for as long as it does not harm the others directly. Of course, this is also associated with the role that liberty has as an ideal, in the Western world and especially in the U.S., and is what really fuels the West-perceived progress and its cultural dominance. Cohesiveness in the society, in the case where cultural norms retreat, is provided by the interdependency of people that is increasing over time. The societies become more and more complex and technocratic, allowing and at the same time also demanding each of its members to possess a domain of expertise. Thus, everyone becomes an indispensable component of the society which in turn, provides a sense of cohesiveness. Of course, interdependency is quite distinct notion from cohesiveness because the former is guided by need while the latter is rather guided by choice. Thus, in order to attain interdependency, the society relies on people's needs that are exploited, compromising their liberty to form cohesive societies by choice.

In biology, we consider as the landmark and a diagnostic characteristic of a species its inability to hybridize with other species, at least to the extent of producing fertile offspring. Two organisms belong to the same species if and only if, as we have learned, they can breed and produce fertile offspring. This requirement was introduced to distinguish between closely related species, such as horses and donkeys, that can breed but any offspring they produce are infertile.

A similar definition, with some limitations, could also be applied to cultural evolution (Creanza et al., 2017). Taking the different cultures as units, in an analogy to the species, they should be regarded as being highly related, to the degree of their identity, only when populations from these cultures can co-exist, interact productively with each other, and generate modes that are highly similar to both of the original cultures that produced them. If the outcome of such interactions is the dominance of one culture over another, then the cultures remain distinct and the culture that is prominent will likely absorb the one that is weaker. An example is provided by the Ancient Greek and Roman cultures. The former fertilized and tremendously influenced the latter, but the Roman culture is seen as distinct from the Greek. Such a definition would also cover cases wherein two distinct cultures co-exist, and merging together to result in a different one, distinct from any of its ancestors. An example of this is the U.S., which is attempting to develop a distinct culture from the amalgamation of the cultures of its various populations. Such an amalgamation sometimes is reflected in new hybrid cultures that remain distant from the pre-existing ones, and sometimes in the co-existence of different cultures in different localities. The culture of Greek–Americans for example, is quite distinct from both that of the Greeks and from mainstream Americans (if such exist); by the same token, we have Chinese–Americans and Korean-Americans who inhabit the various Chinatowns and Koreatowns in the U.S., respectively.

The distinction between a novel culture and the amalgamation of different cultures is often vague. However, the progressive transformation of culture through the mixing of various populations is evident. A more significant diagnostic point may not be the specific ways in which the new culture is recorded in relation to the previous one, but rather the different trajectories it establishes.

For this amalgamation to be productive, it is essential to maintain the link between the cultures and populations; new cultures require diverse populations. When the culture of a single population undergoes change over time, it is still perceived as the same culture that has evolved. For example, Western culture from this standpoint should be viewed as the same from the 17th century to today, despite the substantial differences in ways of life. This continuity exists because the same core ideals have persisted and evolved throughout the years along the same trajectory.

Viewing culture as an evolving entity that is standalone and proceeds independently of the wills of individuals implies that it is subject to growth and changes reminiscent of ontogenetic processes. To appreciate this, it will be useful to engage in a Spenglerian approach and briefly revisit the tradition of Western culture that has resulted in the world we live in today and seems poised to guide our future as well (Spengler 1926). We will identify themes that establish trajectories, making the changes we observe over time more comprehensible. Furthermore, we will appreciate the reasons why culture has, over time, acquired dominance over biology and has become, today, the major, if not the only, force guiding people's lives.

Western civilization: From the Greek civilization to modernity

All societies throughout history have operated by establishing a trade-off whereby individuals forgo some of their liberty to gain access to the benefits of group membership. This activity varied significantly over time and across different regions, and it depended to some extent on the specific civilization in which the corresponding societies were formed. Intellectual frameworks developed to rationalize this trade-off by defining the society's historical and desired needs.

For example, societies that placed a high value on freedom might advocate in parallel that true freedom involves participation in the group and the ability to express one's opinions freely within it, influencing decisions that were to be made. This represented the archetype of the democracy that was developed, matured, and practiced in ancient Athens, which continues to guide the West today.

In contrast, in societies such as those in China and the Eastern Civilization, freedom and liberty were not regard as over-arching ideals. These societies rationalized and advocated for a stricter hierarchical system, emphasizing the Confucian notion that each individual possesses an indispensable, albeit well-structured, role in society. Interestingly, it is worth noting that while in ancient Greece, slavery, the prototypical type of abolition of freedom, was acceptable and widespread, in ancient China was not.

These imaginaries constitute the defining historical and desired needs that people have and contribute the society's cohesion and function (Castoriadis 1987). This reciprocity between what is gained, and what is bestowed by the group's membership reached its maximal breadth in today's Western societies, through the unfolding of the Western Civilization, the one we currently live in, and which appears to be on a trajectory of dominating the world. Today, the maximum number of people within the society are entitled to the benefits of group membership in a way that addresses and resolves both biological needs and those considered essential historical needs for all people, in a non-discriminatory way. Although we have not yet achieved this ideal, it remains a goal to strive for in the future. In the past, and in different contexts, all people had their essential biological needs met, although what was considered minimally acceptable at that time was significantly below today's acceptable standards. Nonetheless, a trajectory has been established, showing that over time, more needs have been satisfied for a greater number of people.

How societies are organized, along with their ideals and value systems, is reflected in their intellectual traditions, which govern the thoughts and concerns of the people. A helpful metaphor for understanding how we reached this stage, as well as for appreciating the trend line we have established, is the analogy between the various major phases of intellectual inquiry in the West and the different stages of cognitive development in humans. Western Civilization traces its roots to the classical Greek period. Since then, different themes have predominated human thought, providing the context for philosophical inquiry and framing the collective mindset of society. Such a framework, defined differently over time, determined the breadth of the group, the fundamental benefits that should be addressed, and the basis by which group membership was justified. They also influenced individuals' expectations from the social hierarchies they developed, reflecting the nature and identity of their historical and desired needs. By monitoring this process, one can appreciate that the progressive transformation of Western civilization across its 2,500 years of history has similarities with the different stages of cognitive development in people, from toddlerhood to adulthood (Kiaris, 2023). According to this scheme, the ancient Greek era, followed by the Middle Ages and then the Enlightenment, resembles toddlerhood, childhood, and the teenage years, respectively. The modern period possesses analogies with human thought at its adulthood.

Let's look at this in greater detail. The ancient Greek Civilization is neither the first nor the only civilization developed by humans since we evolved as a species about 300,000 years ago, and after the agricultural revolution before about 12,000 years. Several different civilizations developed all over the world and eventually, progressively, collapsed. They all developed and practiced agriculture, which triggered the development of hierarchies and complexities in their administration. The unique characteristic of the Greek period was that before them, human thought was predominated by theocratic perceptions according to which life was governed by doctrines imposed by divine entities. Those conveyed the authority to the rulers, instructed the mode of social organization, offered purpose in life, and even rationalized death. All aspects of human existence were determined by boundaries set by the Gods. As such, rigid systems were developed at which the space for intellectual interrogation was rather limited, since the Gods were perfect and therefore, their doctrines would neither needed, and indeed nor even allowed changes. Animism, being the belief that everything in the world has a spiritual essence, should also be viewed as such, from the perspective that it relieves people from the opportunity and responsibility of being able to alter their own lives, in a manner that is systematic and operates at the level of their societies as well.

These aspects all changed with the Greeks, who managed to dissociate intellectual inquiry from theocracy. Philosophy emerged as a discipline in the Hellenic world and eventually evolved to help people address fundamental questions regarding their existence and to answer essential questions about their lives. Philosophical systems and methodologies were developed and continue to guide human thought today. Throughout the years, intellectual inquiry has progressively transformed in terms of the types of questions asked, the context in which they were addressed, the over-arching themes that preoccupied human thought, and the methodologies used to tackle them. These transformations operated reciprocally with human history, reflecting the major stages of human civilization up to the present day. They not only reflected changes but also induced them. By monitoring the main themes of human thought over the years, from the ancient Greek era to modernity, we observe a remarkable similarity to the different stages of cognitive development, from toddlerhood to adulthood. By projecting the lifespan of Western thought onto the life cycle of an individual, we can better appreciate the ontogenetic significance of viewing culture, in this case Western culture, as an entity that has acquired its own trajectory, surpassing the individual people who contributed to it.

From people of faith to people of reason

The Western Civilization traces its roots to the ancient Greek Civilization, during which Western thought established itself. Following a series of transformations, it continues in a rather linear manner until today. The fundamental difference between Greek civilization and all other contemporary, earlier, or even subsequent civilizations that emerged elsewhere was the discovery of philosophy. For the first time in history, humans, by using the philosophical methodology they had established, attempted to answer fundamental questions regarding their existence, lives, and social and political organization, independently of theocratic perceptions and superstitions. Reason substituted faith. Jews, before them, were the people of faith, while the Greeks, soon after the Jews, developed into the people of reason (Barrett 1958). People, alone and independently of the belief in divine interference, established the framework, set the agenda for the questions asked, attempted to answer them, and remained the sole and ultimate beneficiaries of their actions.

They began asking "why"-type questions for the first time, driven by intellectual curiosity rather than by the need to satisfy mere biological demands. Intellect emerged as an indispensable component of their human identity and was seen as equally important, or even more important, than their biological dimension. This perception set in motion the idea that essential needs can extend beyond strictly defined biological needs, and that it is the duty of society and its structures to address them by altering the ways their societies are organized. At this point in history, people understood that their lives operate independently of divine prescriptions, and it remained within their power to define how their lives would be lived.

To achieve this, they engaged in abstract thinking and established a set of attributes for all objects and notions in the world at will, choosing to change them as they deemed necessary. An indispensable component for these achievements was the practice of direct democracy, through which the hierarchical systems they lived in were established by the people. Moreover, these systems could be easily altered if they no longer considered them suitable. The institutions that guided their lives were not beyond criticism and could be compromised if the dialectical process they established suggested such a course of action. By recognizing that every person—every citizen of Athens, at least—had the right and obligation to participate in group decisions through voting, they also acknowledged the direct reciprocity of group membership. This meant that decisions should be made to benefit the group, and through the group, individuals would also receive benefits.

Returning to the analogy of cognitive development, this phase of humanity resembles toddlerhood. According to Piaget (1971), toddlers, from about 2 to 7 years, begin engaging in activities that aim beyond the satisfaction of their biological needs, with pretend play and symbolism being the most prominent. They consciously attribute properties to different objects that they do not possess, and their limits are only their imagination. Initially, their senses were the only means to communicate with the world, but over time they developed language skills. Conversation, which becomes more elaborate over time, extends beyond mere information exchange, establishes a dialectical process, and offers satisfaction as an activity with others. At this stage, they also start asking "why"-type questions, developing reasoning, but it remains mostly simple and immature. Their questions change constantly and are easily satisfied. They do not comprehend hierarchies yet, only the authority of their family, which they accept as it sets their boundaries. Within these boundaries they feel free to act as they wish, as long as they do not present danger to themselves or others. They do not play well with other children at this stage because they do not adhere closely to the rules of others, only those they have established themselves, and they can change these rules at will.

The intellectual stance of the ancient Greeks closely resembles the toddler years of human development. At this stage, the group enjoys full freedom to interpret the world as they wish, to answer any questions they have, and to establish the context in which they live, with their imagination presenting the only limits. No institutions or laws regulate them except those they have established for themselves during that particular period.

In this period, one group of people in particular—the citizens of the city-states—were the ones who enjoyed the benefits of group membership, as the entire society was structured and operated around them. Their most highly-appreciated privilege was their freedom to think and express their ideas, especially in Athens, and to advance these ideas to others if they wished. However, the majority of the population, around 90%, primarily composed of slaves, did not enjoy these privileges. Yet, the existence of this 10% of citizens with access to what society had to offer represents an early, if not the first, genuine attempt of an organized and complex society to offer the privileges of citizenship on a basis that was not hereditary. Importantly, we can also identify another distinctive point: these privileges went beyond the satisfaction of biological needs such as food, shelter, and protection, while they did not aim to satisfy theological concerns by regulating and settling the people's relationship with the divine. In this society, where the Polis included not only the territory but also its people, the identified needs transcended into a different dimension and involved ideals such as intellectual freedom and inquiry, which were considered equally important to biological needs and viewed as essential, deserving protection.

This phase lasted for a few hundred years, during which classic Greek philosophy predominated intellectual life. Over time, both children and humanity begin to comprehend hierarchies and realize that the limits of inquiry are becoming tighter.

Middle Ages and superstition

As humanity entered the Middle Ages, from 300-400 AD to the Renaissance and until about the 15th century, different themes preoccupied human thought compared to the recent past. During this period, the boundaries of intellectual inquiry in the Western world were largely set by the Church, particularly the Western Church, which would later be identified as Catholic. All inquiries of this era operated in relation to theological matters, with the Church determining the limits of intellectual legitimacy. Few questions were posed outside of theology, and even those within these boundaries were answered only in accord with Church doctrines. Consequently, intellectual progress and advancement were minimal during this time, at least in comparison to Ancient Greece or the subsequent Enlightenment, as the boundaries for exploration were too restrictive. The intellectual authority of the Church was beyond question, firmly guiding morality as well. During this period, satisfying theological and religious concerns was an essential component of the people's well-being and reflected crucial historical needs of the time.

This phase is similar to the period of childhood, from about 7 to 11 years. During this time, children further develop their reasoning, although it remains simple and rigid. They seek answers to their questions and prefer simpler explanations, as these align better with their underdeveloped reasoning. At this stage, children fully grasp the hierarchical system imposed by their parents and schools, and they do not question it. Additionally, their age creates firm hierarchical distinctions, where younger or older children represent different groups with lower or higher authority, respectively. Within this rigid system, they may engage in behaviors that could lead to punishment, not out of concrete reasoning to justify their actions but rather because their intuition overrides their fear of consequences. In terms of ethics, they begin to develop a morality based on the codes and norms set by their family, school, and social environment.

The similarity between childhood and intellectual life in the Middle Ages becomes apparent in the fact that both are governed by a hierarchy imposed by the family or the Church, respectively, establishing a system with firm boundaries. Similar to parents who readily satisfy children's questions about the world, the Church's doctrines address the concerns of the people. The process of acquiring knowledge during this period is formal and supervised by family, schools, or the Church. At this stage, systematic knowledge acquisition occurs through educational institutions and the Church, laying the groundwork for the non-incremental progress that follows. Both also implement a firm system of punishments and rewards, which are unquestionable and guide individuals' lives by indicating appropriate and inappropriate behavior. This practice of rewards and punishments establishes paradigms to follow or avoid, contributing to the tight intellectual framework of the period. It also facilitated the functionality of the agricultural feudal societies that emerged, which required adherence to the established hierarchies in order to work.

During this period, both the religious authorities, culminating in the Church, and the state authorities, represented by various kingdoms that developed and persisted over time, viewed it as their obligation to attend to and satisfy the needs of the people. For the Church, these needs were primarily focused on metaphysical and spiritual questions, providing guidance for a religious and Christian life, while state authorities were responsible for meeting demands for protection and sustenance. In both cases, a fairly concrete expectation was established by the people regarding their entitlements from their leaders.

Religion served as an important indicator of group membership, with excommunication from the Church representing a severe form of punishment. Compared to today's standards or those practiced in ancient Greece, the essential needs beyond biological ones were rather limited. It was not until the next phase, particularly during the Enlightenment, that people significantly altered their perceptions to expect additional needs to be treated as essential and thus protected by their leaders. Within the frame of feudal agricultural societies, however, population growth was high, and prolonged periods of stability were experienced.

The age of reason, and treason

As children enter their teenage years, their reasoning becomes more mature. They begin to ask more complex questions centered around philosophical, scientific, and moral themes. Their thinking becomes more abstract, and they tend to idealize reality. The boundaries between plausible and conceivable become vague, which is also connected to their inclination for idealization. During their teenage years, they feel less restricted in making questionable choices, believing there is plenty of time to correct their mistakes. At this stage, they fully comprehend hierarchies and can position themselves within them, albeit in a manner that maintains a critical stance. Although they accept the authority of their families, schools, and social environments, these do not prevent teenagers from criticizing them and following their guidance reluctantly, if they choose to. This reflects their understanding of morality, as they seek to discern the logic behind what is considered moral and what is not, beyond simply being told so. At this point, they also develop their skills and establish preferences that will guide their future lives. They can now envision themselves in adulthood and create conceptual frameworks for their future.

Humanity, during the Renaissance and particularly the Enlightenment, repositioned itself. Superstition was largely abandoned and people began to disengage from the doctrines imposed by the Church. This was the Age of Reason and Science. In the culture that developed, intellectual inquisition gradually liberated itself from theology, and fundamental questions began to be asked from a different perspective, one that does not necessarily involve the divine. Or even if it does so, it remains the active choice of those involved. Fundamental questions are being revisited, such as the essence of human nature, or whether there are other modes of social organization that can be more suitable for people from the existing ones. Unavoidably, such thinking leads to active political discussions and the conceptualization of alternative systems that may have differed drastically from what people had experienced so far. At that time, for the first time after the Greek era, intellectual interrogation again reached a high point and the boundaries were broad, and were accepted as such by the society.

During the end of this period, which lasted almost until the end of the 20th century, plausible political systems spanning the entire ideological spectrum were applied worldwide. Humanity experienced two World Wars with devastating consequences during this time. This was also a period when science and technology advanced at an astonishing pace, laying the groundwork for the industrial revolution of the 18th century, the development of capitalism, and establishing the trajectory that we follow today. Protestant ethics played an important role in these processes by removing the moral stigma from the acquisition of wealth and promoting the values of hard work and simplicity. The Church lost the central role it had held for over 1,000 years and reformed, especially in Northern Europe and the U.S., with the predominance of Protestantism.

The similarities between the teenage years and the period between the Enlightenment and modernity are unmistakable. An over-arching theme in both is the breadth of intellectual inquiry and the conviction—often seen today as a romantic stance—that humanity can reposition itself according to systems based on people's reasoning. Humanity has become increasingly willing to change its condition according to beliefs, guided by the logic that what they view as conceivable is also plausible and ultimately beneficial. The role of superstitions imposed by the Church or established institutions of that period could now be subjected to radical criticism.

Modern period, adulthood, and pragmatism

Following their teenage years, individuals enter adulthood. This is the period during which pragmatism predominates, and the distinction between conceivable and plausible becomes clear. People fully understand that alternative plausible realities may exist, but the only path forward is the one they have already followed, either because it is the most appropriate or the cost to change it is too high to bear. Therefore, while adjustments can be made to improve their situation, radical changes are often avoided. People may not be entirely satisfied with their jobs or the cities they chose to live in, but changing these aspects is not a decision taken lightly.

Pragmatism dominates life in adulthood because, at this stage, individuals must take full responsibility for their lives, as the protective net of family has disintegrated. They must now establish their own protective net to secure themselves and their dependents. Financial matters take precedence in adulthood because they provide a sense of what ultimately matters and finances have a direct impact on people's lives. Intellectual engagement is primarily restricted to the exchange of information that aligns with and reinforces established doctrines, notions, and perceptions. Adults, especially as they grow older, become increasingly unwilling to implement drastic changes in their lives. At this point, they have rationalized hierarchies and aspire to become part of them in a society they feel, in principle, allows them some degree of participation. This is typically achieved through the pursuit of wealth, authority, or a combination of both.

If there is a specific period that could signify the transition of the Western world to its adulthood stage, two key events mark it. The first is the historical collapse of the Eastern bloc. The establishment of communist regimes in Europe during the early to mid-20th century, regardless of how communism was experienced by the citizens of those countries, offered both a conceptual and tangible alternative in the West, facilitating the dialectical process. Critics argued that it was not the system itself that failed but rather its specific implementation in these countries. Consequently, a dialogue was maintained, focusing on historical context, political decisions, and specific leaders that could explain why these systems deviated from the ideological mandates and predictions. However, after the collapse of these regimes in the 1980s and 1990s, a common perception emerged in the West that these ideologies had failed. While a theoretical and academic debate continues regarding the implications of their collapse, for most people, it served as proof that states organized along communist ideals are utopian and destined to fail. Liberal democracy is now celebrated as the sole victor of this process, implying the political future of humanity (Fukuyama 2012). In Europe, where a left-wing tradition remains viable, mainstream progressive political forces have adjusted their arguments to operate primarily as constituents within the existing liberal democratic system rather than offering an alternative path for radically different social organization.

The second event that significantly impacted life at the end of the 20th century is the revolution in information technologies, including the internet and communication. These advancements caused drastic changes that altered the socio-economic landscape, affected work relationships, and raised concerns regarding individual privacy. Furthermore, they fostered a mentality of rationalization and simplicity that reinforces the commitment to technological advancement. The collective conviction is that through scientific and technological research, all humanity's problems will be solved. That belief results in the most prominent characteristic of the present Western society: That the perceived progress itself has emerged as a self-sustained process that operates with its own metrics, occasionally seeming to ignore the well-being of individuals. Funds invested in research from both private and federal sources are constantly increasing, as demanded by society. Frequently, investments are made in new technologies whose actual applications are not yet known, but it is anticipated that solutions will eventually be discovered, highlighting the autonomy with which such progress operates. At this stage, humanity tends to focus on solving problems rather than answering questions. This theme dominates intellectual activity across various domains, reflecting the pragmatism that characterizes the current period. Following a tradition established during the Enlightenment, there is a collective conviction that scientific and technological research will solve all of humanity's problems. This trend suppresses ideological debate and promotes the notion that technocratic solutions are the most appropriate for all challenges.

Another important characteristic of the last few decades is the prominence of environmental concerns and the increased awareness of sustainability issues and the concern over the replenishment of natural resources. These matters are considered important today and that reflects the realism of the present period. This realism is also evident in politics, where efficient management, as opposed to bold ideological stances, holds major importance.

Our adulthood is mirrored in how we view people outside the Western tradition. As adults, we maintain a deep conviction that our ongoing path is the most suitable for humanity, and we have developed a moral obligation to engage all people in this. Therefore, we treat others as our younger brothers and sisters, our dependents, and we proactively export our culture and ideas. We, the Westerners, are the "adults" in this process and the others need our guidance. This cultural export, which contributes to the globalization of cultures around the world, is instrumental for the new dynamics being established in our societies.

The similarities between the adulthood of individuals and the current stage of civilization are evident. In both, realism and pragmatism guide decisions, while the indulgence in drastic thoughts reflects the romantic tendencies of the recent past. Concerns regarding economic matters, expressed as personal income at the individual level or environmental issues at the level of humanity, emerge as priorities that must be addressed urgently. Order in life, society, and humanity represents an ultimately aspired value.

This orientation acquired after modernity resembles one of the two main aspects of the ongoing cultural menopause. The perception that, like older women, Western society has attained an elevated status compared to other cultures, implies a responsibility to train, educate, and transmit available knowledge, facilitating the efficient inclusion of others in the group. Moreover, this must be done impartially and proactively, without considering kinship, cultural, or genetic factors. The other main aspect of menopause, the cessation of reproductive activity, pertains to the biology of contemporary people and will be discussed in a subsequent chapter.

10. Paradoxes of Modernity

The bed of Procrustes

Reaching modernity has enabled us to make a critical evaluation of our current stage, reflecting on whether the goals set have been attained and questioning whether the historical needs established have been fulfilled. Furthermore, it allows for an examination of whether the desired needs we have collectively established, albeit unconsciously and unintentionally, will lead to the improvement of our human condition.

Despite its achievements, Western culture has been associated with a series of problems that seem intrinsic to its progress. Some of these issues stem from internal conflicts arising from opposing forces, while others relate to automation and the demands of sustaining processes that lead to the adoption of goals irrelevant to people's conscious and comprehensible well-being. All, however, represent an unavoidable consequence of the ongoing path we follow. Within this context, the evolutionary fate of humans acquires a new dimension, allowing for a clearer distinction between the needs that were established earlier as biological, historical, or desired. We can recognize signs of this dysfunctionality, where these opposing dynamics meet: it manifests in the gradual loss of individuals' intellectual identity due to enforced conformity, in the conscious abolishment of individuality demanded as a prerequisite for society to function, and in the increasing privatization of individuals that hinders their growth as social beings. The opposing dynamics driving this conflict are the pursuit of humanism and reason—both of which are indispensable ideals of modernity. This conflict places people in Procrustes' bed, where no individual can fit perfectly, being always either too short or too tall. Thus, they are constantly required to either stretch or face amputation.

The progressive dehumanization of society is a consequence of our commitment to technological progress and the conviction that it will readily improve people's lives. Occurring in the West over the last few hundred years, this is probably the first time in human history that society has been structured around the idea of continuous progress, actively pursued at all levels of human life. In this context, progress generates its own dynamics of self-sustainment.

An additional layer to this is the alienation of workers from the products of their labor, which has intensified due to industrialization in modernity. Agriculture is a simpler and more comprehensible process wherein workers maintain a connection with the product of their labor from the initial stages of sowing to the final stages of harvesting and preparation of food or fabric. This process is also relatable to us because of the biological growth it entails. Consequently, the product remains close to the worker and serves as a tangible measure of the success of his efforts, allowing this to be internalized and translated into biological success and population growth.

Industrial production is more complex, and the products are not so obviously the results of one's efforts; they seem artificial. Working in a factory that produces screws, whether for assembling a car or a weapon, one misses the sense of connection that comes from producing food for the table. Therefore, the accomplishment of producing *x* screws cannot be internalized or comprehended as any kind of biological process or success. This distinction will be further explored in relation to demographic trends and the future of humanity.

All this comes as an ironic realization because scientific and technological progress are both direct outcomes of the ideas of the Enlightenment. On one hand, it is widely accepted that science and technology will solve all our problems, thus instructing the establishment of societies driven by technocracy. On the other hand, according to the same Enlightenment ideas, humans should be at the epicenter of intellectual inquiry and the ultimate beneficiaries of its achievements. However, at this point today, the concurrent pursuit of both has resulted in the development of a society that moves forward along a path where humans as individuals have lost their central role. Ironically, this could not have been done otherwise because technocracy displaces people from centrality. Consequently, society is becoming progressively dehumanized, as individuality only acquires value through quantifiable indices aligned with the technocratic trends we have adopted, where success and failure are, and must be, measurable outputs. This trend also instructs an increasingly simplistic perception that transcends human creativity, norms, modes of communication, and even our perceptions of beauty. We gradually abandon elaborate schemes and identify as appealing what is practical and efficient. We also choose increasingly simplistic norms when communicating with others, aiming to convey clear messages that cannot be misinterpreted.

These developments enhance privatization, or withdrawal and isolation, minimizing genuine and more complex interactions that may become critical, restricting them to a foundation that is more formal and stereotypical. Moreover, they amputate individuality in favor of minimalism. An inherent conflict related to this is that people are required to operate under detailed "codes of conduct" that prevail and establish boundaries for appropriate behavior. Beyond legal frameworks, workplaces increasingly dictate to workers which modes of interaction and communication are acceptable, with the power to penalize those who do not conform. It is becoming more common to establish guidelines that describe and prescribe duties, communication modes, and expectations for all aspects of our social and work lives, aligning with the technocratic perception of society, in a context where progress is well-defined and accurately documentable.

On the other hand, the foundation for society's progress within the current system relies on competition and innovation, which compromise the quest for conformity. People are continuously demanded to outperform others, produce more efficiently, and exhibit elements that were not shown before to become high achievers and, therefore, be rewarded by society. To do this, they need to cultivate an innovative and critical stance against what is available and be able to develop something new that outperforms the preexisting. It is this pursuit of the two contrasting tendencies—attaining conformity and seizing the opportunity for innovation—that keeps society in a steady state where tensions are absorbed, and perseverance is warranted. This applies to both individuals as discrete human beings and to humanity as a whole and abstract entity.

These trends dominate various domains of our lives, compromising the values they were developed to nurture. Academic institutions, which traditionally offered a space for intellectual inquiry, provide a typical example. Today, they increasingly emphasize skills development over knowledge cultivation. It is a culture focused on solving problems rather than answering questions. Consequently, the intellectual engagement that once thrived in universities now occurs primarily within specific academic entities or departments, following established methodologies, communicated by experts in the field using specialized technical language and designated media. Intellectual inquiry has diminished its impact on society. This fosters a culture of experts that caters to and fulfills the technological aspirations of society, while minimizing the controversies that can arise from intellectual debates. Ultimately, this reinforces and accelerates the existing trajectory and diminishes individuality.

In the realm of politics, since the Enlightenment, democracy and equality have progressively emerged as the over-arching principles guiding the organization of our societies. Today, however, with the profound impact of technological progress, expertise and technocracy are recognized as the only legitimate values that can provide answers and guide choices. For nearly every question, there exists an expert group capable of expressing an informed opinion and making the right decisions. This expert opinion is valued more highly because it is based on technical data and analyses. Nevertheless, this reliance compromises the principle of equality on which democracy is founded, as not all opinions can carry the same weight. More importantly, the opinions expressed by experts are rarely doubted, given their foundation in technocratic principles that, as derivatives of logic, resist debate.

For strictly technical matters—such as determining what CO_2 levels should be reduced within a given time frame to mitigate global warming, or identifying the most appropriate actions to decrease the impact of a pandemic and prepare for future ones—technocrats are typically seen as the authority on informed opinion. Yet, this unavoidably fosters a culture that permeates all aspects of life, undermining equality and, ultimately, democracy. People tend to accept that others may possess opinions that hold more weight than their own. The emerging culture suggests that actions are taken wisely because reason and logic have formulated the methodologies to address them.

Modernity, a project that can never be completed

The source of this paradoxical state can be traced back to the ideals of the Enlightenment that have guided Modernity since then. On one hand, humanism is characteristically illustrated in Kant's answer to the question "What is Enlightenment?" He stated that "*Enlightenment is the human being's emergence from his self-incurred immaturity...*" and continued by asserting that " '*Have courage to make use of your own understanding' is thus the motto of enlightenment...*" (Kant 1784). On the other hand, the emancipation from superstition by relying on science and reason represented the recommended path to enable individuals to reach their own understanding.

It seems, however, that these two major ideals of humanism and reason are mutually exclusive, and that the project of Modernity is not just unfinished (Habermas, 1981). The strengthening of science and technology, along with their rigidity, means that their conclusions are beyond debate, compromising individual development and progress, and thus one's understanding. Therefore, humanism and reason cannot be reconciled, rendering the project of Modernity unresolvable. This is paradoxical because reason should provide the means to enable humanism conceptually; nonetheless, the culture of technocracy it establishes ultimately suppresses the aspired humanism.

This non-resolvable paradoxical state suggests that it will persist in the future because, by being non-resolvable, it will not allow future stages of cultural development to occur, leaving mankind trapped in this state. Forces aiming to strengthen humanistic ideals will be restricted by the power of reason; whenever humanism recognizes subjectivity and value in individual notions and perspectives, science will advocate for objectivism and linearity. Conversely, the forces that enhance the role of science and technology will be impeded by existing humanistic ideals that demand a repositioning of goals, placing people at the epicenter. This creates an inescapable loop that is further reinforced by the power of the current state to absorb tensions and conflicts. This will prevent crises from occurring, ensuring that the current cultural dynamics remain unthreatened. From this standpoint, we can speculate that the over-arching themes guiding people's lives today will persist in the future as well. Therefore, an examination of humanity today can confidently inform us about the future, at least regarding the factors influencing our genetic heritage. Drawing an analogy to menopause, the future state will be sterile regarding the possibility for radical repositioning, which, in turn, ensures its own autonomous sustainability and perseverance.

11. Selection and its elimination

Reproductive inequality in *H. sapiens*

Unavoidably, cultural changes find their effects in human biology. Gene frequencies can change from one generation to another in a group because some individuals have greater reproductive success than others, leading to their descendants and genes being overrepresented in the population over time. This is not necessarily because certain genes are better suited for a given environment and thus positively selected, while others are negatively selected. It can simply result from the more intense reproduction of some individuals, which may be due to resource availability or other social reasons. For several mammals, including non-human primates, their status in the group determines their reproductive success: the higher their dominance and status, the more females are available for reproduction. This results in what has been designated as reproductive skew and reflects the reproductive success of certain individuals within the group. If this dynamic changes over generations, it may not lead to evolutionary changes. However, if it persists and remains steady in specific lineages, it may consistently alter the gene pool.

What about humans? A study conducted over 20 years ago evaluated the genetic legacy of Genghis Khan, the formidable leader of the Mongols who lived almost a millennium ago, in today's populations (Zerjal et al., 2003). DNA analyses provided clear evidence that about 8% of the population in regions across Asia, from the Caspian Sea to the Pacific, carries identical stretches of Y chromosome. This corresponds to approximately 16 million people or 0.5% of the total population and can only be explained by selection that occurred in the male lineage. Likely, this reflects the genetic legacy of Genghis Khan in today's population and serves as an example of how culture and dominance, at its most unequivocal manifestation, can impact the genetic landscape of people. Since the times of Genghis Khan, several changes have occurred in our societies, and differences in status do exist, although prestige has now substituted dominance in acquiring status.

People are very sensitive to understanding differences in the social status of others. Not only are the obvious cues for demonstrating wealth informative in this aspect, but also simple cues related to even brief speech that can indicate class and guide hiring decisions (Kraus et al., 2019). Thus, societies remain highly sensitive in interpreting signals that inform about social class and integrate them into their choices.

Several studies have been conducted, across different populations corresponding to various cultures and societies, systematically addressing the extent of reproductive inequality and measuring reproductive success in relation to status and other factors (Ross et al. 2023; von Ruedena & Jaeggi, 2016). Generally, the results of these analyses confirm a skew, where certain males exhibit higher reproductive success than others, associated with their status. However, this skew is much lower than that recorded for other primates. One factor that reduced its impact compared to other mammalian species is the acquisition of monogamy, which inhibits the direct translation of status into reproductive performance, optimizing it among different males in the group. Monogamy is imposed with varying degrees of stringency but is prevalent in various societies across the majority of human populations. In some cases, it is favored or promoted, while in others it is strictly enforced. Monogamy serves to reduce competition among males, thereby lowering the projection of potential differences in prestige, status, genetic fitness, and reproductive efficiency. This cultural choice to favor monogamy diminishes the genetic impact of another cultural choice that favors the acquisition of prestige and status.

The impact of monogamy in depressing reproductive skew is not exclusive to human populations; it is also evident in other monogamous species, including certain rodents like the California mouse (*Peromyscus californicus*), where male reproductive skew is comparatively low relative to similar species. As discussed earlier, monogamy is linked to a series of evolutionary adaptations such as parental care, which are common in monogamous mammals. In these species, mating partners form lifelong bonds and typically mate with another partner only after losing their original mate and usually after a certain period. Such binary partnerships benefit not only the social strategies they promote but also the health and well-being of individuals, as the disruption of these pair bonds can lead to detrimental health consequences. Laboratory disruptions of mating pairs render animals more susceptible to serious pathologies, including cancer (Naderi et al., 2021; Ford & Young, 2021).

Monogamy ultimately stimulated a more egalitarian organization in the societies of our ancestors, where more benefits are realized by identifying as group members rather than as individuals. Consequently, our emerging sociality reflects not only our need for interaction and satisfaction from such engagements but also serves as an ecological strategy to collectively enhance our fitness by becoming group members. Monogamy mitigates the negative consequences of competition in a society that increasingly values prestige in status acquisition. Thus, by dissociating reproductive performance from status, it aids in separating culture from the biological trajectory of our species.

A series of evolutionary adaptations have evolved in parallel, providing a clear physiological foundation for the pleasure and satisfaction derived from our social interactions, which promote such interactions (Krach et al., 2010). From that perspective it is our biology that is instructing us to form and maintain relations with others because it regulates our nervous system in a manner that elicits the sense of pleasure. Consequently, as the network of these interactions becomes more complex, the group tends to grow larger, and different strategies are developed and optimized. These strategies may occasionally appear to be drastically opposed to one another, leading culture to be perceived as distinct from biology (Whiten et al., 2017). Nonetheless, all these elements were triggered by our biology and find their foundation in our biological substance, making it very difficult to determine where one ends and the other begins.

Coming back to the original question regarding the existence of reproductive inequality among people, we find that today we exist in a group larger than ever before, with a common denominator in the continuously expanding Western culture. This culture defines the framework of our lives and offers us an identity. Our shared culture enables us to communicate and interact with one another, even with individuals physically distant from us, with whom we may not even speak the same language. In this case, the group is defined not only by geographical borders but also by its expanding common culture. More intriguingly, the assimilation into Western culture has emerged as a desired need for many individuals in non-Western cultures, guiding their political landscapes. For example, Turkey, situated at the intersection of East and West since the times of Kemal Atatürk in the early

20th century, has consistently had a major political party advocating for the country's Westernization. Similarly, in China, during the later stages of the Qing dynasty and especially in the early 20th century, this mindset in favor of Westernization was expressed through the New Culture Movement. In Iran, the social forces that supported the Shah in the second half of the 20th century, which remain viable today in both Iran and especially among Iranians abroad, were also in favor of Westernizing Iranian society. In the West, however, while social forces exist that are suspicious of the ongoing path, they do not seek to join another cultural trajectory. Instead, they advocate for the radical or less radical transformation of the existing path towards an alternative trajectory that still constitutes a subsequent step in the trajectory established by Western culture. Thus, our current system of ideals as Westerners, imposed upon us by the societies in which we were born and raised, reinforces our common identity and appears to be the prevailing one. Through this system of ideals, we are encouraged to identify ourselves and attribute meaning to our lives.

The transition from farming to industrialization is typically associated with demographic transitions, during which the association between wealth and fertility, seen previously, becomes weaker or dampened (Pérusse, 1993; Skirbekk, 2008; Vining, 1986; Shenk et al., 2016). While science is making strides in eliminating biology-driven inequality, cultural and socioeconomic factors still influence whether we will have children, when we will have them, and how many children we will have, in ways that neither biology nor the availability of resources can strictly limit. The decision of whether and when to start a family is conscious and is heavily impacted by the priorities imposed by the sociocultural environment.

The demographic transition refers to the shift from high birth and death rates typically seen in agricultural societies to lower birth and death rates that accompany industrialization (Bongaarts 2009). The reduction in death rates precedes the reduction in birth rates; thus, in the initial phase, the population increases until it is balanced by the subsequent decrease in birth rates. These demographic changes take over a century to conclude and vary in their phases and intensity across different populations. Cumulatively, they reflect the net changes occurring in populations, but different groups of people, depending on their societal status, respond differently to the socioeconomic changes that take place.

Analyses of demographic data may provide insights into these questions. The first question is whether inequalities in wealth and income impact people's reproductive patterns and decisions. The *Gini* index reflects economic inequalities across different countries, with a value of 0 indicating that wealth is evenly distributed among individuals and a value of 1 reflecting maximum inequality, where all income goes to one individual and none to another. A study exploring the association between reproductive age, or age at marriage, and the *Gini* index showed a significant positive correlation. In countries with high inequality, people tend to start their families later, suggesting that living in societies with high income inequality shifts individuals' priorities toward personal growth and development rather than starting a family (Cheng et al., 2020). From this perspective, we already see a dissociation of individual interests from family.

Hopcroft (2018) analyzed demographic data in U.S. populations from the Survey of Income and Program Participation to explore potential associations between the childbearing partners that men or women have in relation to their income and net worth. Education showed a negative association with the number of childbearing partners in both sexes, reinforcing the link between monogamy, culture, and education. Net worth and income were positive indicators for the number of childbearing partners and fertility in men but negative for women. More detailed analyses indicated that this positive correlation in men is primarily due to a lack of childbearing partnerships when their income is very low, rather than a higher number of childbearing partners among high-income individuals. Finally, when childbearing unions are multiple, it results in more children when income is very high or very low than when income is in the middle range, for both men and women.

These data convey significant information, clearly showing that socioeconomic factors remain a source of reproductive inequality in the Western world. Furthermore, they highlight differences between women and men in how they prioritize the choice to engage in a childbearing partnership, depending on the resources available to them. Notably, both very high and very low incomes lead to more children, likely because in these extreme cases, the allocation of resources among the children is less of a concern—either due to abundance or due to scarcity.

To fully appreciate the contribution of resource inequality to reproductive inequality, one must also consider the positive feedback dynamics that emerge. Single parenthood, particularly motherhood, is increasingly common and has a racial component, being more frequent among African American women. The burden of supporting a child is greater for single mothers, exacerbated by the fact that African American women face higher financial resource inequalities. Thus, the resulting reproductive inequalities intensify, fueled by the ongoing disparities in resource availability for specific cohorts of the population (Baker 2018; McLanahan & Percheski, 2008).

These are only a few of the several studies available that provide abundant data on the differences in reproductive success associated with socioeconomic status. The results become more complicated when non-trivial relationships are unveiled, such as the higher number of children among those with very high or very low incomes. Thus, certain groups of people will produce more children than others for a variety of reasons and a combination of reasons.

Childlessness is another trend that characterizes modern societies. From about 3.5 children per mother in the 1800s, today, we see fewer than 2 children per mother, which is also associated with a progressive increase in single-parent households over time (Stone 2020). Family structures are changing, with fewer children per household, fewer households with both parents present, fewer marriages, and increasing childlessness.

Cultural traditions, family structure, how existing resources are allocated to children, and even the significance of bearing children in modern societies account for these trends. Nonetheless, in the absence of tangible associations between wealth and the genetic make-up of individuals, such inequalities will reproduce themselves, albeit socially, but are not expected to drive biological changes or alter the gene pool of populations. Wealth and resources will vary among people, but gene frequencies will not follow these changes.

Projecting reproductive inequality in the future

How will reproductive inequality evolve in the future? Is it going to increase? Will it be eliminated? Or will it acquire a different profile from what exists today while still affecting the genetic makeup of future generations? Let's discuss some individual factors that can affect it, and then we may be able to make some predictions. First, we should distinguish between biological factors that directly influence the fertility status of prospective parents and the fitness of children, and those that indirectly impact reproductive success through social constructs. In both cases, we need to keep in mind that what matters for evolution is reproductive success, not the quality of life of individuals. The latter matters only if it affects people's reproductive success and their offspring's chances of survival; otherwise, it is irrelevant. If wealthy and successful couples consistently produce two offspring, while couples with fewer accomplishments and poorer conditions produce five, regardless of the quality of life and future opportunities, the latter couple will be more successful from a biological genetic perspective because they will have more children. Their genes will be over-represented in the genetic pool of the population. If the offspring of poorer couples remain poor and continue to have more children than those of the wealthy couples for several generations, then a genetic trend emerges whereby the genes of poorer lineage will increase in frequency within the population. The question then shifts to whether these two types of couples, the poor and the rich, have differences in their genes. The definitive answer is no; they do not, provided that the distribution of wealth and its maintenance from one generation to the next does not follow genetic underpinnings. Furthermore, it assumes that social mobility is eliminated to prevent the mixing up of genes between groups of people that possess different wealth status. Otherwise, not only are wealth and status determined by the genes, but also the groups of people with differences in wealth and status consistently intermarry within their group. Therefore, no significant interference in the gene pool will occur.

For the first category, which includes factors impacting reproductive success in the narrowest biological sense, we can anticipate the near-elimination of factors affecting reproductive efficiency and fertility in the future. This is a direct consequence of recent advances in medicine and associated technologies. Over the past few decades, remarkable progress has been made to address all sources that could reduce fertility, reaching a point today where almost everyone can have children. Assisted reproduction, which emerged a few decades ago, has enabled couples who otherwise could not have children to do so, and technologies like in vitro fertilization have helped both men and women become biological parents. Furthermore, women who may carry genes that could potentially reduce their lifespan can cryopreserve oocytes while still healthy and use them in the future, even if they become ill or receive treatment. Looking ahead, tangible opportunities are emerging, restricted now primarily because of regulatory laws, to produce offspring by manipulating somatic cells, enabling everyone, regardless of sexual combinations, to have children. Active research in this direction promises to realize these efforts, dissociating the genetic continuation of one's lineage from the traditional physical reproduction methods seen in mammals.

The impact of this is dual: First, it assists individuals who are practically infertile to become fertile, enabling them to transmit their genes to the next generation. Second, in cases where couples cannot have children, the cause often resides with either the man or the woman; however, the other partner might still be able to have children with another partner. Thus, biologically, infertility affects the individual but it also impacts their partner through social forces, unless they choose to mate with someone else. In such cases, not only is a single lineage affected but rather two, preventing certain combinations of genes from coming together—those corresponding to each partner. Assisted reproduction eliminates this barrier, increasing the pool of potential gene combinations and facilitating the concept of random mating, hence allowing all gene combinations to occur within a population.

Limitations in offspring production and their fitness do not arise solely from infertility. Tremendous effort is currently directed toward providing everyone the opportunity to lead a life as normal as possible, despite conditions that previously posed strict limitations. For example, individuals living with sickle cell disease, if properly managed, can experience a normal pregnancy today. Furthermore, it is likely that in the future, infertility—a common complication in males with this condition—will also be addressed (Gallo et al., 2010; Jain et al., 2010). Women diagnosed with cancer can preserve their oocytes for use after completing therapy (Fraison et al., 2003; Mahajan 2015). Similar progress is evident in the management of infectious diseases, where the discovery of antibiotics and recent treatment advancements have significantly reduced their toll on the population. A relevant example is the recent COVID-19 pandemic, which, had it occurred 100 years ago, would likely have claimed millions of lives.

More importantly, in the case of COVID-19, and likely other pandemics as well, the population that appears most vulnerable is the elderly, thereby reducing the genetic burden in the population. Although we are still a long way from managing such pandemics as effectively as we would like, our efforts are directed toward this goal, and it seems we are on the right track, particularly if we focus on the prospective success in enabling everyone born to become capable of producing a biological child.

What is even more astonishing is not just the medical and technological progress that allows such opportunities, but also the way society embraces these advancements to enable those in need to take advantage of them. Provisions are being made to allow individuals with disadvantages to lead normal lives and reproduce. It is not only medical progress that aids this, but also the extensive social network established to care for children, support them, and undertake the obligation of their upbringing. While society has not yet fully resolved these issues, a path is in place that receives the support of the community as a whole.

From a biological perspective, the collective goal of our society, supported by the state, charity, or other structures we have developed, is to offer every individual the opportunity to produce offspring and subsequently support them until they reach reproductive age. In other words, our collective goal is to maintain the existing gene pool while dissociating it from biology and its impact on performance. Differences in reproductive success among individuals, which are strongly associated with specific socio-economic statuses, may justify reproductive inequalities; however, these inequalities will not systematically affect the population's gene pool because there is no tangible association between genetic make-up and socio-economic status. Furthermore, social mobility, another key goal of our prevailing culture, will not allow such inequalities to persist in the long run for specific genetic lineages. Although social mobility has not increased but rather declined in the last century in the U.S., the rise in income and resource availability above minimum levels does not interfere with reproductive dynamics (Song et al., 2020). Gene frequencies may change over time in parallel with the societal status of the individuals carrying them, but these two forces—the change in gene frequencies and the change in individuals' status—will operate independently of each other, without any intention of favoring one over the other. From a genetic perspective, both will be random and unrelated: the status of individuals will not be reflected in their genes, and the genes that prevail or replenish will not do so based on the information they carry.

12. The Future of humans

Dynamics of future evolutionary changes

With these trends in mind, we can speculate on the framework within which our society will operate in the future, particularly regarding the factors that can impact our biology. The key elements contributing to the shaping of this emerging landscape should consider advances in medical and other technologies, the group sizes in which future populations will reside, the humanism that emerges after modernity, and finally, the occurrence of unpredictable, random events that can affect human populations and their gene frequencies. The first three factors have already been discussed from the perspective of the adaptive effects that certain genes may have and their dissociation from their impact on the human gene pool. The fourth factor remains the only one capable of driving changes in our biological future as a species, systematically or abruptly altering gene frequencies.

1. Medical and other technological advances will enable everyone to bear children and thrive in an environment where external factors will not impose limitations. This includes technologies that assist reproduction and those that allow individuals to reach reproductive age. Even anticipated inventions aimed at enhancing our physical performance—collectively described as contributors to our new phase as metahumans—will not alter the gene pool, as they will function within a society where reproductive performance is dissociated from personal well-being. We may be able to dictate text to our computers using our thoughts or incorporate

hardware into our bodies to enhance strength or efficiency, but this will not influence the number of children we have. If it does, we will have to assume that fundamental ideals that currently guide us will be abandoned, a possibility that is rather unlikely for reasons we discussed elsewhere.

2. An increase in the size of the group to which one belongs. From this perspective, borders between different populations become increasingly porous, leading to unification within the group one identifies with. The mixing of genes and diverse genetic combinations, which once seemed unlikely, will occur with greater probability now that cultural and geographical borders are collapsing.

3. The dominance of Western culture and its ideals. This interacts with the previous factors, highlighting the growing perception and capacity of society to support all individuals within an expanding group. Furthermore, it signifies the belief that everyone should be regarded as a legitimate recipient of the benefits that society can offer.

4. Changes in gene frequencies can occur due to factors other than biological selection. Events such as large-scale catastrophes that eliminate significant portions of the population, as well as random occurrences like genetic drift and gene flow, will continue to influence gene frequencies in ways that remain fundamentally random.

The dominance of randomness

The first three factors are the ones that, combined, dampen selective pressure by eliminating the adaptive value of genes from reproductive performance. The fourth element remains the only factor outside of the lab, capable of driving future evolutionary changes in humans. Gene frequencies will not remain stable over time. For gene frequencies to remain stable in a population, certain conditions must be met. Some of these are being fulfilled, while others are not.

First, this stability assumes that the population is very large. This condition materializes over time due to the continuous expansion of the group and the elimination of geographical, and more importantly, cultural borders.

Second, it presupposes that each individual will have equal chances to mate with every other individual in the group. Currently, this is not the case because, although the group expands, cultural and socio-economic factors, or even geographical distance, continuously alter the probabilities of mating. First, the two individuals must physically meet to start a family together. Second, guided by their own mind sets, they need to recognize in someone a "worthy" prospective mating partner. Cultural and socio-economic factors will interfere, reflecting preferences that can promote or inhibit certain choices. However, increasing social and physical mobility will prevent specific trends and preferences from stabilizing across generations. Society, through changing policies and redefined desired needs, eliminates trends that could stabilize at a level affecting gene frequencies in the long run. Thus, such choices will remain random at the genetic level.

Finally, it presupposes that all matings will result in the same number of children, so no specific genes will gain an advantage by being overrepresented in the next generation. Childless marriages and differences in the number of children among different groups will ensure that this condition will never be satisfied. Therefore, gene frequencies will change over time and from one generation to another, as well as among different localities or groups of people with varying socio-economic statuses. Nonetheless, due largely to the Westernization of our emerging global culture, none of these changes will be attributed to the information that the corresponding genes carry. Even attractiveness, which can drive the selection of a mating partner and has the potential to influence selection—especially when other forces are absent—is highly subjective and unstable as a selectable trait. Studies comparing facial images have shown high variation in attractiveness based on both the culture of the individual in the photograph and the culture of the evaluator (Voegeli et al., 2021). Even greater diversity in the ideal of beauty exists when viewed through the lens of history (Romm, 1987; Dimitrov & Kroumpouzos, 2023). For biology, such changes will remain random; even if they may transiently favor specific lineages, their inconsistency over time would prevent them from establishing persistent trajectories.

Another factor that may, in the future, alter the frequencies of genes in our collective gene pool is the occurrence of abrupt, catastrophic events that can result in the sudden death of many individuals. Such events instantly eliminate corresponding genes from the gene pool, and since the distribution of genes will never be uniform within a population, the representation of these genes in the remaining population will change. Massive earthquakes, pandemics, global wars, or even an asteroid impact are examples of such events. Doomsday scenarios that have the potential to drastically reduce large fractions of the existing population will always be possible. Our science and technology aim to predict such occurrences to mitigate their catastrophic consequences or to manage them after they happen and minimize their impact. Nonetheless, their occurrence cannot be ruled out and, practically, in the long run, will not be averted. Such events will again be random from a biological perspective because they will not target specific groups of people based on their genetic makeup. In that sense, the toll of these events will be borne evenly by the population, irrespective of individual genetics. The remaining population will be different, but this change will be random and not associated with the event that caused it.

One exception by which external changes may trigger alterations in the gene pool in a manner that targets specific genes is likely that of pandemics. Although medical management occurs on a genetically neutral basis, different individuals may have varying sensitivities based on the genes they carry. However, to accept the possibility of an effect occurring and having evolutionary consequences, we must assume that the surviving population will have a different genetic makeup, sufficiently enough to change our evolutionary fate. From the trajectory we have established, aiming to contain and eliminate the impact of such pandemics with potential evolutionary significance, they should also be considered random events. A pandemic that kills individuals carrying allele A1 more frequently than those with A2 will result in the enrichment of the A2 allele. It is possible that this occurrence will simultaneously decrease the frequency of allele B1 compared to B2, simply because A1 individuals happen to have B1 at a higher frequency at the time of the pandemic. If B1 is associated with a human trait, its depletion from the gene pool and the enrichment of B2 may lead to evolutionary changes. Yet, since it is not the one targeted by selection and is only indirectly affected, such change will be random. The same occurs if A1, targeted by the pandemic, also happens to be associated with another human trait alongside sensitivity to the pandemic agent: its change will also induce evolutionary changes, but those will be irrelevant to adaptation.

An analogy can be seen in our views regarding plausible environmental events that may harm some individuals more than others, based on the specific genes they carry, thereby creating disparities. Our assumption is that, in the long run, we will be able to address such events to eliminate these disparities. The reason we can accomplish this, sooner or later, is that such events, if they target people depending on our specific biological identity, will be scientifically predictable and manageable. We will address these issues through our efforts to eliminate disparities, which is an over-arching ideal of our humanism. Therefore, they provide opportunities to tackle them with our science, as opposed to events that harm irrespective of our specific biology and remain random.

Another plausible event that could impact our evolutionary trajectory is the colonization of other planets. This concept sounded like science fiction until recently, but today it seems increasingly likely, with systematic efforts to colonize Mars, for example, underway. In this scenario, we can speculate that a small number of people will be the colonists, from which a new population will be established. Founder effects will be unavoidable because this group of colonists, by definition, will not have a gene pool representative of the one they originated from. Even if they do, their evolution will not follow the same genetic trajectory as the parental population, due to drift and randomness.

In the long run, such founder effects may result in considerable genetic changes, leading to a population with a distinct profile compared to the parental one, or in other words, in evolution. This should occur only if gene flow from the parental population is restricted to nullify ongoing changes. The fact is that such events will be random and irrelevant to the impact these genes have on individuals' reproductive performance. If this new colony retains the same culture as the one from which it originated, the ideals of the Western tradition—a rather likely possibility considering that colonization is Western trait—will likely prevent discrimination and ensure that individual capacities are not limited or restricted. Considering that this population will be relatively small in size, it is plausible that such evolutionary changes will occur more rapidly than in the parental one.

The collective brain hypothesis, which is largely responsible for our human identity and cultural evolution, has established an evolutionary trajectory that may guide future evolutionary changes for our species (Dunbar 2009; Gavrilets 2015; Schimmelpfennig et al., 2022). According to this hypothesis, human evolution is driven by the evolution of the brain, which has tripled in size and become increasingly capable of storing, handling, and benefiting from large amounts of information (Muthukrishna et al., 2018). An intriguing observation that supports the evolutionary significance of brain size in information storage is that, over the last 3,000 years, human brain size has actually decreased (DeSilva et al., 2021). While this could simply correlate with a decrease in body size, this reduction in brain size is about five times less pronounced than the decrease in body size, and it occurred earlier. Thus, the likelier explanation is that brains became smaller because social complexity and group decision-making increased and became more sophisticated, relieving the body from the burden of sustaining this energy-demanding organ.

The enhanced ability of people to cooperate with each other was essential for our cultural evolution and formed the basis for the development of social norms grounded in shared moral codes (Boyd & Richerson, 2009). Several theoretical approaches, based on both modeling and experimental analyses, demonstrate the advantages of collaborative performance and cooperation, explaining why these behaviors persisted and strengthened throughout our evolutionary history (Bowles & Gintis, 2011; Ridley, 1998). However, despite these advantages, it is challenging to accept that cooperation will impact our evolutionary future, as such benefits cannot be translated into the reproductive success of specific individuals and of their lineages within our established societies. Even if groups of individuals with enhanced collaborative capacities secure better access to resources than those without, thus potentially increasing their reproductive performance, our social norms should inhibit these capacities from affecting the reproductive success of their offspring, preventing them from becoming selectable traits. Our culture ensures that such benefits become accessible to everyone by promoting their impartial distribution throughout society. This impartiality, which is essential in our current and likely future norms, eliminates the selectability of these traits.

Forecasting the future

Humans, like all other species, evolved under the influence of a combination of factors, with selection always playing a significant role in enabling them to adapt more efficiently than their predecessors. Among their other characteristics, they developed intellect, which in turn allowed them to foster culture. Culture also evolved and was transmitted through generations, guiding their lives and establishing the value systems that shaped their societies. At various points in their history, humans have placed a special emphasis on caring for every member of their group, irrespective of individual capacities and biological fitness. An initial examination of this attribute could be explained in terms of evolutionary theory, to the extent that considerable genetic similarity can be identified between those who provide assistance and allocate a fraction or all of their resources and their beneficiaries. Nonetheless, this trend, exaggerated by science and technology and reinforced by the cultural ideals of Western society, especially in the modern period, has reached a point that challenges evolutionary theory, as the groups to which individuals identify themselves are now much broader than ever before and independent of genetic relevance.

The dissociation of fertility from status and resource abundance intensified in modernity, as reflected in the ongoing demographic transition, where the high birth and high death rates of agricultural societies are followed by low birth and low death rates after industrialization (Bongaarts, 2009). In agricultural and pre-agricultural societies, fertility was generally associated with status (Skirbekk 2008; Colleran et al., 2015). In its initial phase, the demographic transition typically involves a decrease in death rates due to advances in healthcare and technology, which precede the decrease in birth rates. This decrease in birth rates eventually offsets or even exceeds the birth rates, stabilizing or reducing population size. The reduction of birth rates in industrial societies is puzzling because status does not correlate with reproductive efficiency (Vining 1986). Initially, before the new humanism of the Enlightenment became fixed in the collective mindset, population growth after industrialization was primarily governed by the reduction of death rates. Subsequently, despite continued resource increases, birth rates decreased because well-being became dissociated from reproduction. Echoing Marx, the alienation of people from the products of their labor in the industrial setting may contribute to this trend of reduced birth rates. As people become estranged from the products of their labor and the wealth they generate, they cannot intuitively internalize it and translate it into resources that maintain the biological link between population growth and resource abundance. In agricultural societies, this link remains more viable, and families grow larger due to higher birth rates when available resources permit it. In these societies, people maintain a viable connection with their work products, making it easier to translate abundant resources into traditional biological outcomes. A potential underlying reason for this is that agricultural production is simpler than industrial, making it more comprehensible to workers. It involves a product that biologically grows, allowing for a natural connection with the process and its outcomes. Participation in this process is voluntary, as each step can be comprehended and performed by every worker, from the earliest to the latest stages. In contrast, industrial production's scale and complexity prevent people from comprehending it in its entirety, leading to alienation. As a result, the products of their labor cannot readily be internalized and translated into the biological success of collective labor. Various issues in modern industrial society, such as depression and behavioral disorders, may be linked to this disconnect between biological success and work effort (Obschonka et al., 2018; Hidaka, 2012). This is also relevant to the fact that people today do perform work, in the form of hobbies, that resembles production as it was performed in agricultural and pre-agricultural societies, but not work that characterizes industrial societies. For example we enjoy cultivating the land, gardening, or becoming amateur carpenters, but we do not voluntarily engage in processes that simulate industrial production. Apparently, we receive a sense of satisfaction from the one but not the other.

In the future, especially in the West, as industrialization is progressively replaced by information technologies in the workforce, this alienation will likely intensify, with more pronounced consequences for people than what Marx described for the industrial societies of the 19th century. The labor of an information-based workforce not only translates into products that are foreign to the workers, but often tangible products do not exist at all. Consequently, the disconnection between workers and their labor will become more significant, leading to more severe consequences for population dynamics and perceptions of biological success. From this perspective, the demographic paradox will intensify.

It is challenging, and probably irrelevant, to identify the causality in these events—whether the reduction of birth rates followed resource abundance due to a higher investment by people in their new social and work roles or preceded it as a prerequisite for engaging in industrialization. Nonetheless, it aligns well with the new position people have acquired in Western societies.

Culture now dictates that reproductive performance is unlinked to status and biology, asserting that every human deserves the benefits of progress and care, irrespective of whether they are part of the cultural group that can provide aid. Furthermore, everyone should receive the benefits of progress and have the opportunity to reach reproductive age and have children. In this case, the collective goal of our society is to maintain the existing gene pool and ensure that it persists in future generations.

This reflects the second similarity our current state as a species has with postmenopausal women. The first similarity is the acquisition of the status of teachers and imitation models directed towards all cultures and people in need, through means unrelated to genetic relevance. Additionally, provisions have been made to eliminate possibilities for radical change and transition into alternative models that can govern our society. This second similarity relates to the loss of reproductive capacity. By actively engaging in a trajectory to maintain our genetic pool—not only ours but also that of other species—we commit to stabilizing our genetic make-up, which is equivalent to preventing evolutionary changes that might otherwise occur in the future. Therefore, this postmenopausal phase relates to both our cultural and biological attributes.

However, *H. sapiens* will not cease to evolve; it will continue because genes will be transmitted and will be present in different combinations and frequencies among individuals, collectively modifying the phenotypes of their carriers. The difference, however, will be that the forces driving evolution will change. The ideals established through our culture prohibit the differences in new phenotypes that emerge from being translated into adaptive differences. Even if new traits arise that benefit certain individuals, they will not translate into better reproductive efficiency because our culture will strive to optimize reproductive efficiency for every group member and eliminate disparities. Thus, selective advantage will not translate into evolutionary outcomes.

Evolution will be driven by phenomena that are random in nature. Most importantly, genetic drift—which changes gene frequencies in populations over time, independent of individual costs and benefits—will be a driver of evolutionary change. This will also be enhanced by partner preferences within the group based on existing inequalities, which will likely persist in future societies. In this context, since genetic drift is higher when populations are smaller, we would expect that greater inequalities and associated partner preferences would strengthen genetic drift. A prerequisite is that such partner preferences and inequalities persist across several generations for the same groups of people and their descendants. However, since these inequalities will never be associated with the individuals' genetic make-up from a biological perspective, even if they consistently affect genetic frequencies, they will result in random effects.

Evolution in controlled environments

To understand how evolution might proceed in the absence of selection, we can revisit an example from small mammals maintained as closed colonies in animal facilities. Different stocks of deer mice, the small rodents mentioned earlier in our discussion on monogamous behavior, have been kept in captivity for several decades, with the oldest stocks dating back to the 1960s (Havighorst et al, 2017). A colony manager selects different animals as breeders approximately every two years to maintain the stocks, and the colony continues through these randomly selected breeders. We can estimate that around 30-35 generations have passed since the original inception of the stocks from their wild-type ancestors. In human terms, assuming each generation spans about 25 years, this corresponds to roughly 800-900 years, equivalent to the time since the High Middle Ages. Under these conditions, selection has been eliminated from the deer mice because food is abundant, so the animals do not need to secure their food independently, and even the choice of mating partners is determined by the colony manager. Only unavoidable selection for the best breeders persists since animals that produce very few offspring are less likely to have their genes over-represented in the next generation, as the following generation of breeders will likely not be genetically associated with the poor breeders.

Furthermore, sick animals and their genes are also purged from the colony because they are less likely to be represented in the next generation of breeders. Surprisingly, after almost 900 human years in captivity and under no selection, the fundamental characteristics of these species—such as monogamous behavior, anxiety due to partner loss, exploratory behavior, and reproductive performance—remain almost intact over the years (Havighorst et al, 2017). Monogamous species still behave monogamously, they grieve when they lose their partner, and they show parental care, while non-monogamous species maintain their polygamous behavior despite the forced monogamy over 30 generations (Naderi et al., 2021). Thus, what is genetically ingrained in these species does not change, even when the environmental factors that originally induced and stabilized it are lost. One might argue that 900 human years is too short a time for evolutionary changes to be detectable in such multifactorial behavioral characteristics. This argument holds some validity, but it suggests that humans, over the years in the absence of selective pressure, will remain as they are today, even if environmental conditions change drastically. No matter how people evolve in the future or the challenges they face as individuals and as societies, their behavioral traits are unlikely to change significantly due to external factors. The interesting point is that this lack of selection in people is not the result of a colony manager but rather the effect of the culture we have developed and strengthened, which effectively eliminates such pressure. It appears that our culture serves as an analogy to the colony manager of the deer mice and the controlled animal facility.

Other phenomena that can drive evolutionary change include events related to sudden changes in population size. Catastrophic events that can eliminate large fractions of the population, such as wars, environmental disasters, or pandemics, as well as founder effects linked to the potential colonization of other planets, may also drive evolution. However, these events will also be random from a genetic perspective, as they will not relate to the genetics of those affected. Thus, it seems that the evolution of *H. sapiens* will continue in the future, but for the first time in biological history, it will be based on randomness. Since our culture has dissociated adaptation from reproductive success, we have actively set the goal of prohibiting evolutionary change. This ensures that *H. sapiens* will be the first and only highly intelligent species ever to orient its collective resources to halt its evolutionary change. However, since our evolution cannot be completely stopped, the new species that may emerge from us in the future will evolve guided by randomness, irrespective of their biological fitness. This realization is paradoxical; our advanced intellect has led to this outcome. Our intelligence and behavior, which have meticulously cultivated and advanced our science and technology through reason, have succeeded in disrupting the evolutionary logic of biology, eliminating the power of selection as a driver of human evolutionary change, and putting an end to a process that began millions of years ago when biological time commenced. Currently, as a species, we make every effort to ensure our biological perseverance while unintentionally allowing any changes that may occur to be based exclusively on randomness. We have not yet achieved this, as several of our biological limits regarding reproductive success still persist, but this is the trajectory we have established for our biological future. In that case, we can speculate that the new species that will replace *H. sapiens* may possess better or worse fitness than us in biological terms, simply because we, through our culture, will not allow changes that impact our fitness to establish themselves in the population and become dominant. Society, through its collective ideals and priorities, will actively attempt to nullify the consequences of both biological disadvantages and advantages. This dissociation of fitness from reproductive efficiency allows only random events to guide future developments.

In this new realm guided by randomness, humans may become shorter or taller, darker or lighter in color, more or less intelligent, and they will remain as such, without compromising our collective species identity. Existing genes, along with new ones and their combinations that may enter the gene pool, will be prevented from acquiring any advantage that can allow them to predominate. This is not because they will be specifically targeted, but rather because the remaining genes will persist and be supported by our social norms.

The only deviation from this trajectory could arise from the collective abandonment of the ideals that currently predominate, particularly those advocating for impartiality. If this occurs, we could predict a society that would disregard the well-being of all its members, potentially favoring certain groups at the expense of others. Ultimately, this would affect the allocation of resources and, consequently, their chances for reproduction, leading to new dynamics in our evolution. For this to happen, a drastic change in current social norms is required, to an extent that sufficiently alters our cultural trajectories. Such a deviation is the only factor that can compromise our culture, which views the entire species as a group. This perspective allows us to examine the voices that challenge the ideals that guide us. These voices often promote notions of partiality in care, sometimes driven by an overwhelming support for the ideal of liberty at the expense of equality. By abandoning our collective conscience, which recognizes the benefits of our sociality, we inadvertently allow biological evolution to affect humans by reinstating the role of selection. However, this presents a circular argument, as our sociality is an inherent feature of our species, and the prevailing dynamics will fortunately prevent such changes from occurring.

13. An Aristotelian ramification with the defeat of the unmoved mover

Aristotle, Darwin, and teleology

Unavoidably, contemplations regarding the evolutionary future of our species carry repercussions in philosophical thought and should reference Darwin and his evolutionary theory. They should also refer to Plato and especially Aristotle, who established the trajectory of Western thought, which in turn resulted in today's society and its traditions. To that end, concerns regarding teleology and its absence resurface.

A point of contrast between Aristotle's and Darwin's views of life is related to the steady perception of species in Aristotle, as opposed to their changing nature according to Darwin. Not surprisingly, Darwin developed his theory of evolution through natural selection by examining wild species of plants and animals. Aristotle acknowledged a stability in species description, proposing that they differ merely due to their different forms. He also established the philosophical trajectories we discussed earlier, according to which humans, through reason—an original Aristotelian attribute—brought evolution to an end.

Nonetheless, the perceived biological "steadiness" in Aristotle may resemble Darwinian change when the concepts of potentiality (Δύναμις) and actuality (Ενέργεια) are integrated into evolutionary thinking. In Aristotelian cosmology, a key notion lies in the principles of potentiality and actuality (Bodnar 2018; White 2020). The first represents the various possibilities a thing can acquire, while the second reflects the fulfillment of potentiality when it becomes materialized through a broadly defined movement or process. For example, the potentiality of wood is to become a table, through the action of the carpenter that provides the actuality. Similarly, in the living world, a seed possesses the potentiality of developing into a tree, with the actualization of this potentiality conferred by the biological substance of the seed that can grow and change.

In this cosmology, motion is mediated by an indefinite number of movers that are also moved by another mover, generating an infinite number of interlinked potentialities and actualities. Nevertheless, according to this line of thinking, it becomes evident that there must be a single mover that remains unmoved, responsible for the motion of all other movers and the motions they induce. This unmoved mover, according to Aristotle, culminates in the ultimate purpose that lacks matter and therefore potentiality, possessing only pure actuality (Bodnar 2018). This entity offers cause and purpose and is equivalent to a divine being, as it is devoid of any potentiality (Merlan 1946; Olson 2013). This implies the typical Aristotelian teleology, which presupposes a goal and purpose in everything, established in advance, well before its materialization. In this teleological scheme, actuality is prior to potentiality, and life is organized hierarchically from plants to animals, to humans, and ultimately to God. In this strictly hierarchical structure, mind and reason progressively acquire increasing significance and role, starting with plants that lack reason and culminating in God, who embodies both actuality and reason.

From the Darwinian standpoint, organisms evolve continuously, adapting to environmental changes by means of natural selection (Darwin 1859). This is an ongoing process going back to when life began. As such, it lacks teleological considerations and thus, Aristotelian hierarchies may become debatable (Solinas 2015; Lennox 1993; Godfrey-Smith, 1993). Teleology, however, is recognized by several thinkers in Darwin, so we may consider this debate ongoing (Ayala 1970; Gotthelf 1987; Lennox 2001). Indeed, as discussed by Boucher (2021), such a debate regarding the role of teleology in Darwin's thought began in his own time and continues until today. Boucher (2021) convincingly proposes, though, that such disputes are merely verbal and not substantive, while Kampourakis (2020) dissociated the design teleology or intention, or need-based, from selection teleology or the one that is based on a natural process, and accepted as valid only the latter.

The discussion of teleology in Darwin's thought, and in biological evolution in general, primarily concentrates on the effects and contextualization of biological functions and their association with traits and designs. However, as regards the evolutionary process as a whole, especially in the context of a hierarchy that identifies an ultimate point, teleology is rejected because evolution is open-ended, it is unpredictable, and it is non-linear. This is because of the randomness that is instrumental in evolutionary thought. Evolution is ongoing, bearing no substantive arguments that it will ever stop.

Evolution of humans and the impact of culture

While the open-ended and ensuing nature of evolution is a valid notion when it refers to all species that ever evolved, there is a big caveat in the direct application of this evolutionary process to humans. As we discussed extensively before, through the technological and scientific progresses that the humans achieved, we developed a highly anthropocentric culture which defeats the operation of evolutionary logic, as this operates in all other species. This accomplishment largely relies on the Aristotelian principles of logic and reason that provides the foundation for modern science and technology. Our cultural evolution has established norms that essentially prevented evolutionary change in the biology of humans, at least at the extent that it is induced by natural selection. Such norms are the consequence of our cultural evolution that intensifies altruistic behavior and expands group selection to a scale that now as a group, the whole species of *H. sapiens* is included (Boyd & Richerson, 2009; Wilson & Wilson 2007).

A humanistic culture has been established, especially after the Enlightenment, in which reproductive success, for the first time in biological history, is not associated directly with the individuals' biological fitness. Success in our society is not measured by the offspring produced, relieving the fundamental Darwinian entanglement of reproductive success with selection. This is "revolutionary" from the evolutionary perspective since always, reproductive efficiency, at the level of the individual or the group, is the one that measures fitness. To that end, the evolution of the human species will stop being the outcome of natural selection, since our egalitarian values, combined with the opportunities offered by medicine and technology, prevent the Darwinian "survival of the fittest" from driving the evolutionary progress. Although selection of such egalitarian culture, from the perspective of group selection at its widest scale, may be viewed as a character that can be selected favorably in the future, the fact that it is applicable to, and targets all humans, is making it to cancel out itself as a selectable character. This refers to the collective brain hypothesis according to which, since egalitarianism is intended to apply to all, irrespective of the altruistic intuitions of the individuals, it will persist but will not be selected against potential non-egalitarian groups and behaviors. To favor partially the egalitarians would contrast its egalitarian nature.

Importantly, these shifts in the evolutionary framework of humans have been performed consciously by our species, by a process that is based on our Reason, which in Aristotelian cosmology is upwards in the teleological hierarchy (Sedley 1991; Capecci 1979). From the evolutionary perspective we can see such progress in science, technology, and medicine, being integrated in a manner at which the pressures driving natural selection are being eliminated from operating as evolutionary forces.

Evolution and Aristotelian cosmology

If we project the evolutionary changes that have occurred over time onto Aristotelian cosmology, we can parallel species' evolution to a progressive transformation driven by changing potentialities through natural selection. From this perspective, all species that have ever existed can be viewed as potentialities that have been actualized, with natural selection serving as the major force driving evolution. In this case, natural selection provides the actuality through which continuously changing species acquire their potentialities over evolutionary time. This applies to all species that have ever existed, constantly changing and being replaced by others better adapted to their historical conditions. For some extinct species, this linear process has ceased, while for others, which we consider ancient remnants today, it may be progressing too slowly. However, the overall scheme remains valid.

The controversy and the irony

By accepting the aforementioned reasoning and by focusing on the humans and their evolutionary lineage, we will come to the recognition of an apparent controversy that is also accompanied by an ironic realization:

The controversy is that the scientific and technological progress, that rely on Aristotelian logic, have defeated the evolutionary logic that operates in all other species, stopping the evolution of humans, at least through classical natural selection. As instructed by our active choice as a species following the onset of human civilization, we are consciously attempting to diminish the impact of selection, dissociating it from the evolutionary dynamics of our species. Furthermore, we have rendered as our collective goal as a species, our perseverance over time.

The subsequent ironic realization is that by deliberately halting our evolution (which confers actuality) through our egalitarian culture and the progress enabling it, we reject future unrealized potentialities, rendering us the ultimate point in the Aristotelian hierarchy. In this case, in the absence of actuality, our species embodies a pure potentiality. Such potentiality can conceivably change, but will not, since the actuality that could have altered it has been eliminated. Therefore, while approaching Aristotle's unmoved mover by enhancing our actuality through reason and logic over evolutionary time, we are concluding our evolutionary journey by becoming a potentiality discordant to any actuality. As a result, Aristotelian teleology is being reinstated; despite the increasing significance of logic and rationale for our species, our ultimate point—the so-called Entelechy or Telos—exists as a state of pure potentiality, as the actuality of natural selection has been intentionally eliminated.

Considering that stochastic events, such as genetic drift, mutations, isolation, catastrophic doomsday events, and major pandemics, will continue to act as evolutionary forces, randomness is now gaining enhanced significance as the driver of human evolution. From an opposite perspective, it is also important to appreciate that the underlying force of these transitions is the emergence of culture, which occupies a central position that drives changes irrespective of the dynamics generated by biology, narrowly viewed. Thus culture, instead of evolution, may confer the new actuality that persists independently of its carriers, the human beings as biological entities.

This recalls the previously discussed paradox of modernity, by which humanism and reason are inherently incompatible. In the Aristotelian scheme, the ultimate abolition of actuality, through the process of its continuous dominance, has resulted in a rigid state of potentiality devoid of reason. This state reflects the current situation where culture, by dominating our biological substance, restricts reason, even as it actively promotes it. Humanism, on the other hand, from the perspective of enabling our own understanding, which must rely on reason to advance, is lost in the context of a culture that promotes rigidity in its goals and aspirations in a self-sustained process.

14. Biological existentialism

Culture in humans extends an extra layer of plasticity to our biological substance that is absent in animals and other species, where the biological essence is tighter, more rigid, and instructive. What we perceive as human nature, broadly defined, changes over time according to cultural context because, at the level of our existence, we cannot separate the two. Needs and priorities differ at various times but are always perceived as essential and instructive. This is reflected in what we are willing to sacrifice at different times to gain what we consider essential for our well-being. It is also reflected in what people living today in different or even the same society are willing to sacrifice for something else. In both cases, time and cultural context can drastically redefine the worthiness of our sacrifices.

Although the biological foundation remains constant, the cultural component being nurtured does not; yet we internalize it at any given time as having the same significance and impact as the biological. It is continuously invented throughout our history as a species and is defined differently depending on the historical period. At any given time we are subjects of years of collective history that produces a unique result, consisting of a stable biological component and a changing cultural component. From that perspective, culture provides a dynamic entity that persists over time and influences the substance of humans. Thus, while for all other animals their essence (or substance) pre-exists their existence, humans, by incorporating the cultural component, manage to have their existence precede their biological substance. Priorities traditionally seen as biological recede in favor of cultural imperatives due to our strengthened sociality and interdependence, making culture a more stable and powerful constituent in how we perceive, interpret, and evaluate our existence over time.

However, since our cultural attributes can only be defined in reference to us, human beings, biology should also be accorded priority. Therefore, the two components—one corresponding to nature and the other to what can be nurtured—should not be seen as distinct but rather as intermingling, acquiring priority only from the specific perspective by which they are examined and how the expected benefit is defined.

When viewed together, the cultural and biological perspectives suggest that both as a species and as cultural and intellectual products, humans have reached a distinctive and unique point in their history. Regarding their biological substance, we have defeated the evolutionary logic guided by the selection of the fittest, leaving their future path to randomness. What will happen to mankind in the future does not rely on the evolutionary logic that shaped its path so far. It will largely be determined by random factors that may or may not align with the logic of evolution. This traditional notion of evolution presupposes that each species is more efficiently adapted than its predecessor, thus replacing it. Culture, especially the culture of care and empathy, which instructs the expansion of the group beyond narrowly defined and genetically explained altruism, coincided with the evolution of *H. sapiens* and is taking on a global perspective today. Such culture effectively counteracts potential deficits in biological fitness due to the advantages offered by science and technology, a direct product of our culture.

The case is similar regarding humanity's intellectual journey, which has now reached a point recognizing that a concrete value system has been developed. We have reached a point where we feel that we have established modes of organization that can persist over time. As a system also based on modernity's logic and reason—an indispensable component of our cultural tradition—these values should hold forever, at least as long as our scientific method remains the same. This is also a system that is resilient and flexible, capable of accommodating pressures and changes.

The major challenge that still remains is to apply this system to all of humanity and modify it incrementally over time, improving it as needed. We collectively believe that drastic changes in our stance are not likely, as the system we have proudly developed, advocated, and supported should be able to absorb tensions and integrate conflicts. Not only our moral values but also our entire imaginary have acquired a robust position, because logic and technology are viewed as avenues that will provide solutions to every problem that eventually emerges. In that sense, they ensure stability based on the products of the scientific method and technology, which are indisputable in time and space.

Within this context, humanity is entering an existential crisis. Both intellectually and biologically, it seems we are at the end of a given path, after which our direction will not be guided by dynamics based on any type of logic but rather on the flux of randomness. Our scientific and technological achievements can now absorb biological deficits. In this context, humanity as a group, and individuals as entities, will need to identify purpose and invent goals and aspirations to infuse meaning into their existence. Such meaning can be dissociated from biological imperatives, as repeatedly demonstrated in our history, according to which the imperatives dictated by our culture have already been prioritized over those dictated by our biology.

The fact is that all this relates to, and can be effectively supported by our culture. Care and empathy have defeated the traditional logic of biology that predominated during our evolutionary journey until now. In the absence of a collective culture, evolutionary logic remains the major force that can manifest itself, prioritizing our biological substance.

Culture is an inherent component of our human existence and not an incidental invention; as such, it has now acquired self-fulfilling dynamics. In line with the famous quote from Sartre (2007) that "*existence precedes essence*," existence can define essence and, therefore, resolves the dichotomy between culture (or nurture) and biology by identifying it as a tautology. From this perspective, aspects of our behavior and culture that might have been seen as contradictory to our biology now simply reflect the increasing breadth that biology acquires. This represents a goal of modern biology: its effort to widen its realm by identifying biological substance in phenomena traditionally studied by other disciplines, such as psychology, behavior, and sociology.

From the opposite perspective, we can recognize a dichotomy between nurture and nature, but such a dichotomy is only superficial because they both represent two inseparable components of the human condition. History is the process through which humans attempt to expand the component of nurture at the expense of nature. Success and progress are measured as the defeat of nature and are evaluated by these accomplishments. States and societies, historical and otherwise, where nature emerges as more dominant than nurture are considered "primitive." The same process of "nurturing," through education for example or the establishment of cultural norms that protect individuals from the deficits their own nature provides, is viewed as a reflection of cultural advancement. This also applies to all human activities attained without obvious biological benefit, which are perceived as reflections of cultural advancement. Even going back to the archetypical cultural reflections, such as rituals associated with death, the inception of the divine, and the establishment of religious practices and philosophies, they are also unique to humans.

We are the only species that creates artifacts that do not fulfill a direct biological purpose. Through our culture, and in a way that is very subjective across different cultures, we attribute properties to these artifacts that remain beyond our comprehension and worship them. This power of culture illustrates its autonomy as a process that surpasses its originating essence as a human construct. From this standpoint, culture acquires the power to govern our biological substance and progressively displace it. Such artifacts, both material and noetic, are products of our culture, surpass our individual substance, and acquire a self-fulfilling essence. They change over time, but any given period they set the boundaries within which we perceive our existence.

Through culture, we are able to redefine our essence since it is preceded by our existence. Our biological substance becomes malleable and not limiting because of our technological advancement and the culture that dictates its use. Without culture, biological essence precedes existence, defining limits and restricting priorities, as in all other species, where existence is defined by essence. Culture has defined a new existence, prior to and independently of biological essence. Humans come into being in a predefined cultural context that is carried and developed throughout the years, changing at every different time.

The so-called biological or evolutionary logic, applicable to all species but difficult to apply directly to humans due to cultural interferences, can now be re-examined by considering this logic as part of a unique substance that integrates culture as a component. Our cultural identity already exists when we are born, as our biological essence comes into being. As such, it has the capacity to define different forms of essence manifested differently at different times. Furthermore, these forms can also be manifested differently according to our will, increasing the plasticity of our biological substance. We can decide if and when to have children, how many, and to what extent our resources will be committed to them. We can also determine the magnitude of our interference with other species and other people. Such decisions are influenced by our current culture but remain intentional at the scale of mankind viewed as an entity.

In today's world, culture has given us unprecedentedly broad margins within which we can define our essence. This is a liberation of human beings, as we can now appreciate the malleability of our existence and transcend our biological imperatives, thanks to the cultural heritage we have established and continue to strengthen. At the same time, however, we must adhere to our cultural imperatives, the persistence of which depends on the inherent conflict they sustain: to promote both reason and humanism simultaneously. The first strengthens a unified perception of our existence in the collectivistic context of the group as the most efficient strategy to move forward, while the second fosters the individualistic stance that enables personal growth and provides the means for appreciating the fulfillment of such growth. It will remain challenging for us, though, as our conscience originates and concludes at the individual level, making us vulnerable to the perception that cultural imperatives are more external than individual. This represents the existential crisis we face when we discover the absurdity that the cultural constituent contributes to our lives. This realization sustains a sense of randomness, leaving us with only one viable option: to embrace it.

Bibliogaphy

Adhikari R. Bringing an end to deadly "menstrual huts" is proving difficult in Nepal BMJ 2020; 368 :m536 doi:10.1136/bmj.m536

Adhiyaman V, Adhiyaman S, Sundaram R. The Lazarus phenomenon. J R Soc Med. 2007 Dec;100(12):552-7. doi: 10.1177/014107680710001201З. PMID: 18065707; PMCID: PMC2121643.

Adhikari K, Chacón-Duque JC, Mendoza-Revilla J, Fuentes-Guajardo M, Ruiz-Linares A. The Genetic Diversity of the Americas. Annu Rev Genomics Hum Genet. 2017 Aug 31;18:277-296. doi: 10.1146/annurev-genom-083115-022331. PMID: 28859572.

Albert, David Z., 1992, Quantum Mechanics and Experience, Cambridge, MA: Harvard University Press.

Ainsworth, Thomas, "Form vs. Matter", The Stanford Encyclopedia of Philosophy (Summer 2020 Edition), Edward N. Zalta (ed.), URL = ‹https://plato.stanford.edu/archives/sum2020/entries/form-matter/›.

Assis, S., A. L. Santos, and C. A. Roberts. 2011. "Evidence of Hypertrophic Osteoarthropathy in Individuals from the Coimbra Skeletal Identified Collection (Portugal)." International Journal of Paleopathology 1 (3): 155–163. doi:10.1016/j.ijpp.2012.02.009.

Austad SN. Menopause: an evolutionary perspective. Exp Gerontol. 1994 May-Aug;29(3-4):255-63. doi: 10.1016/0531-5565(94)90005-1. PMID: 7925746.

Ayala F. Teleological explanations in biology. Philosophy of Science, 37 (1970), pp. 1-15.

Baker, R. S. (2022). Ethno-Racial Variation in Single Motherhood Prevalences and Penalties for Child Poverty in the United States, 1995–2018. The ANNALS of the American Academy of Political and Social Science, 702(1), 20-36. https://doi.org/10.1177/00027162221120759.

Barrett, W. (1958). Irrational man: A study in existential philosophy. Doubleday & Co.

Basow SA, Howe KG. Role-model influence: effects of sex and sex-role attitude in college students. Psychology of Women Quarterly. 1980;4:558–72.

Binder M, "Jaspers and Ortega on the Historicity of Being Human," Existenz 14/1 (2019), 28-34

Bodnar, Istvan, "Aristotle's Natural Philosophy", The Stanford Encyclopedia of Philosophy (Spring 2018 Edition), Edward N. Zalta (ed.), URL = ‹https://plato.stanford.edu/archives/spr2018/entries/aristotle-natphil/›.

Bongaarts J. Human population growth and the demographic transition. Philos Trans R Soc Lond B Biol Sci. 2009 Oct 27;364(1532):2985-90. doi: 10.1098/rstb.2009.0137. PMID: 19770150; PMCID: PMC2781829.

Bocquet-Appel JP. When the world's population took off: the springboard of the Neolithic Demographic Transition. Science. 2011 Jul 29;333(6042):560-1. doi: 10.1126/science.1208880

Bonné-Tamir B, Nystuen A, Seroussi E, Kalinsky H, Kwitek-Black AE, Korostishevsky M, Adato A, Sheffield VC. Usher syndrome in the Samaritans: strengths and limitations of using inbred isolated populations to identify genes causing recessive disorders. Am J Phys Anthropol. 1997 Oct;104(2):193-200. doi: 10.1002/(SICI)1096-8644(199710)104:2‹193::AID-AJPA5›3.0.CO;2-#. PMID: 9386826.

Bowles, S., & Gintis, H. (2011). A cooperative species: Human reciprocity and its evolution. Princeton University Press.

Boucher, S.C. Biological Teleology, Reductionism, and Verbal Disputes. Found Sci 26, 859–880 (2021). https://doi.org/10.1007/s10699-020-09728-3

Boyd R, Richerson PJ. Culture and the evolution of human cooperation. Philos Trans R Soc Lond B Biol Sci. 2009;364(1533):3281-3288. doi:10.1098/rstb.2009.0134

Buss DM, Schmitt DP. Mate Preferences and Their Behavioral Manifestations. Annu Rev Psychol. 2019 Jan 4;70:77-110. doi: 10.1146/annurev-psych-010418-103408.

Butticè, Claudio (2019). Universal Health Care. Santa Barbara, California: Greenwood Publishing Group. pp. 20, 25. ISBN 978-1-4408-6844-3.

Capecci A. (1979) Final Causality and Teleological System in Aristotle. In: Tymieniecka AT. (eds) The Teleologies in Husserlian Phenomenology. Analecta Husserliana (The Yearbook of Phenomenological Research), vol 9. Springer, Dordrecht. https://doi.org/10.1007/978-94-009-9437-9_2

Castoriadis, Cornelius. The Imaginary Institution of Society. Trans. Kathleen Blamey. Cambridge: MIT Press, 1987.

Cheng L, Zhou X, Wang F, Hao M. The greater the economic inequality, the later people have children: The association between economic inequality and reproductive timing. Scand J Psychol. 2020 Jun;61(3):450-459. doi: 10.1111/sjop.12612. Epub 2020

Cilliers L, Retief FP. The evolution of the hospital from antiquity to the end of the middle ages. Curationis. 2002 Nov;25(4):60-6. doi: 10.4102/curationis.v25i4.806

Colleran H, Jasienska G, Nenko I, Galbarczyk A, Mace R. Fertility decline and the changing dynamics of wealth, status and inequality. Proc Biol Sci. 2015 May 7;282(1806):20150287. doi: 10.1098/rspb.2015.0287

Creanza N, Kolodny O, Feldman MW. Cultural evolutionary theory: How culture evolves and why it matters. Proc Natl Acad Sci U S A. 2017 Jul 25;114(30):7782-7789. doi: 10.1073/pnas.1620732114. Epub 2017 Jul 24. PMID: 28739941; PMCID: PMC5544263.

Cressey, Paul F. "The Influence of the Literary Examination System on the Development of Chinese Civilization." American Journal of Sociology, vol. 35, no. 2, 1929, pp. 250–62. JSTOR, http://www.jstor.org/stable/2766126.

D'Amato FR. Effects of male social status on reproductive success and on behavior in mice (Mus musculus). J Comp Psychol. 1988 Jun;102(2):146-151. doi: 10.1037/0735-7036.102.2.146. PMID: 3396312.

Darwin, Charles, 1809-1882. On the Origin of Species by Means of Natural Selection, or Preservation of Favoured Races in the Struggle for Life. London :John Murray, 1859.

Dawkins, R. (1976). The selfish gene. New York, Oxford University Press.

DeSilva Jeremy M., Traniello James F. A., Claxton Alexander G., Fannin Luke D. When and Why Did Human Brains Decrease in Size? A New Change-Point Analysis and Insights From Brain Evolution in Ants. Frontiers in Ecology and Evolution (9) 2021. 10.3389/fevo.2021.742639

Diamond, J., Rotter, J. Observing the founder effect in human evolution. Nature 329, 105–106 (1987). https://doi.org/10.1038/329105a0

Dickel DN, Doran GH. Severe neural tube defect syndrome from the Early Archaic of Florida. Am J Phys Anthropol. 1989 Nov;80(3):325-34. doi: 10.1002/ajpa.1330800306. PMID: 2686462

Dimitrov D, Kroumpouzos G. Beauty perception: A historical and contemporary review. Clin Dermatol. 2023 Jan-Feb;41(1):33-40. doi: 10.1016/j.clindermatol.2023.02.006.

Drake JW. The distribution of rates of spontaneous mutation over viruses, prokaryotes, and eukaryotes. Ann N Y Acad Sci. 1999 May 18;870:100-7. doi: 10.1111/j.1749-6632.1999.tb08870.x. PMID: 10415476.

Duguid S, Wyman E, Bullinger AF, Herfurth-Majstorovic K, Tomasello M. Coordination strategies of chimpanzees and human children in a Stag Hunt game. Proc Biol Sci. 2014 Dec 7;281(1796):20141973. doi: 10.1098/rspb.2014.1973. PMID: 25320165; PMCID: PMC4213656.

Dunbar RI. The social brain hypothesis and its implications for social evolution. Ann Hum Biol. 2009 Sep-Oct;36(5):562-72. doi: 10.1080/03014460902960289

Durkheim, E. 1963. [1898] La prohibition de l'inceste et ses origines. L'Année Sociologique 1: 1-70. Reprinted as Incest. The nature and origin of the taboo, trans. E. Sagarin. New York: Stuart.

Ellis L. 1995. Dominance and reproductive success among nonhuman animals—a cross-species comparison. Ethol. Sociobiol. 16, 257–333. (10.1016/0162-3095(95)00050-U)

Flint, M. (1975). The menopause: Reward or punishment? Psychosomatics, 16, 161-163

Ford CL, Young LJ. Harnessing the healing power of love. Trends Mol Med. 2021 Sep;27(9):833-834. doi: 10.1016/j.molmed.2021.07.010. Epub 2021 Aug 4. PMID: 34364786; PMCID: PMC8605763.

Fowler, J. and Christakis, N., 2009. Connected: The Surprising Power of Our Social Networks and How They Shape Our Lives. New York: Little, Brown and Co.

Fowler, C., Olalde, I., Cummings, V. et al. A high-resolution picture of kinship practices in an Early Neolithic tomb. Nature 601, 584–587 (2022). https://doi.org/10.1038/s41586-021-04241-4

Fowler J. H., Settle J. E., & Christakis N. A. (2011). Correlated genotypes in friendship networks. Proceedings of the National Academy of Sciences of the United States of America, 108(5), 1993–1997. https://doi.org/10.1073/pnas.1011687108

Fraison E, S Huberlant, E Labrune, M Cavalieri, M Montagut, F Brugnon, B Courbiere, Live birth rate after female fertility preservation for cancer or haematopoietic stem cell transplantation: a systematic review and meta-analysis of the three main techniques; embryo, oocyte and ovarian tissue cryopreservation, Human Reproduction, Volume 38, Issue 3, March 2023, Pages 489–502, https://doi.org/10.1093/humrep/deac249

Frankopan P. (2019). The New Silk Roads: The Present and Future of the World. New York : Alfred A. Knopf, 285 p.

Fry DP and Söderberg P. Lethal Aggression in Mobile Forager Bands and Implications for the Origins of War. Science 341, 270 (2013);DOI: 10.1126/science.1235675

Fukuyama, Francis. 2012. The End of History and the Last Man. Harlow, England: Penguin Books.

Gallo AM, Wilkie D, Suarez M, Labotka R, Molokie R, Thompson A, Hershberger P, Johnson B. Reproductive decisions in people with sickle cell disease or sickle cell trait. West J Nurs Res. 2010 Dec;32(8):1073-90. doi: 10.1177/0193945910371482. Epub 2010 Aug 11. PMID: 20702680; PMCID: PMC3590021.

Gao, J., Su, Y., Tomonaga, M. et al. Learning the rules of the rock–paper–scissors game: chimpanzees versus children. Primates 59, 7–17 (2018). https://doi.org/10.1007/s10329-017-0620-0

Gardner A, West SA. Social evolution: the decline and fall of genetic kin recognition. Curr Biol. 2007 Sep 18;17(18):R810-2. doi: 10.1016/j.cub.2007.07.030. PMID: 17878052.

Garg S, Anand T. Menstruation related myths in India: strategies for combating it. J Family Med Prim Care. 2015 Apr-Jun;4(2):184-6. doi: 10.4103/2249-4863.154627. PMID: 25949964; PMCID: PMC4408698.

Gavrilets S. Collective action and the collaborative brain. J R Soc Interface. 2015 Jan 6;12(102):20141067. doi: 10.1098/rsif.2014.1067. PMID: 25551149; PMCID: PMC4277098.

Gignoux CR, Henn BM, Mountain JL. Rapid, global demographic expansions after the origins of agriculture. Proc Natl Acad Sci U S A. 2011 Apr 12;108(15):6044-9. doi: 10.1073/pnas.0914274108.

Gilbert LA, Gallessich JM, Evans SL. Sex of faculty role model and students' self-perceptions of competency. Sex Roles. 1983;9:597–607.

Godfrey-Smith, P. (1993). Functions: Consensus without unity. Pacific Philosophical Quarterly, 74, 196–208.

Goldschmidt, Asaf. Reacting to Epidemics: The Innovative Imperial Public Health System during the Late Northern Song Dynasty. Chinese Medicine and Culture 6(1):p 68-75, March 2023. | DOI: 10.1097/MC9.0000000000000041

Gottlieb A. Menstrual Taboos: Moving Beyond the Curse. 2020 Jul 25. In: Bobel C, Winkler IT, Fahs B, et al., editors. The Palgrave Handbook of Critical Menstruation Studies [Internet]. Singapore: Palgrave Macmillan; 2020. Chapter 14.

Gracia A, Arsuaga JL, Martínez I, Lorenzo C, Carretero JM, Bermúdez de Castro JM, Carbonell E. Craniosynostosis in the Middle Pleistocene human Cranium 14 from the Sima de los Huesos, Atapuerca, Spain. Proc Natl Acad Sci U S A. 2009 Apr 21;106(16):6573-8. doi: 10.1073/pnas.0900965106

Graeber David and Wengrow David. The Dawn of Everything: A New History of Humanity. Farrar, Straus and Giroux, 704pp, 2021

Green JP, Holmes AM, Davidson AJ, Paterson S, Stockley P, Beynon RJ, Hurst JL. The Genetic Basis of Kin Recognition in a Cooperatively Breeding Mammal. Curr Biol. 2015 Oct 19;25(20):2631-41. doi: 10.1016/j.cub.2015.08.045. Epub 2015 Sep 24. PMID: 26412134.

Greshko M and Natl Geo Staff 2019. https://www.nationalgeographic.com/science/article/mass-extinction.

Guterman MA, Mehta P, Gibbs MS. Menstrual taboos among major religions. The Internet Journal of Wealth and Societal Politics. 2007;5(2):1–7.

Habermas, J. 1981. 'Modernity versus Postmodernity'. New German Critique

Haig, D. (2020). From Darwin to Derrida: selfish genes, social selves, and the meanings of life. Cambridge, Massachusetts, The MIT Press.

Haldane J. B. S. The rate of mutation of human genes. In: Proceedings of the Eighth International Congress of Genetics. Hereditas 35, 267–273 (1949).

Hamilton WD. The genetical evolution of social behaviour. Journal of Theoretical Biology 1964 7 1 52.

Hardy, K. et al. Neanderthal medics? Evidence for food, cooking, and medicinal plants entrapped in dental calculus. Naturwissenschaften 99, 617–626 (2012).

Havighorst A, Crossland J, Kiaris H. Peromyscus as a model of human disease. Semin Cell Dev Biol. 2017 Jan;61:150-155. doi: 10.1016/j.semcdb.2016.06.020.

Hawkes K, O'Connell JF, Jones NG, Alvarez H, Charnov EL. Grandmothering, menopause, and the evolution of human life histories. Proc Natl Acad Sci U S A. 1998 Feb 3;95(3):1336-9. doi: 10.1073/pnas.95.3.1336. PMID: 9448332; PMCID: PMC18762.

Henrich J. 2015. The Secret of Our Success: How Culture Is Driving Human Evolution, Domesticating Our Species and Making Us Smarter. Princeton, NJ: Princeton Univ. Press

Henrich J, Gil-White FJ. The evolution of prestige: freely conferred deference as a mechanism for enhancing the benefits of cultural transmission. Evol Hum Behav. 2001 May;22(3):165-196. doi: 10.1016/s1090-5138(00)00071-4. PMID: 11384884.

Herrmann E, Haux LM, Zeidler H, Engelmann JM. Human children but not chimpanzees make irrational decisions driven by social comparison. Proc Biol Sci. 2019 Jan 16;286(1894):20182228. doi: 10.1098/rspb.2018.2228. PMID: 30963858; PMCID: PMC6367165.

Herrmann E, J Call, MV Hernàndez-Lloreda, B Hare, M Tomasello, Humans have evolved specialized skills of social cognition: The cultural intelligence hypothesis. Science 317, 1360–1366 (2007).

Hidaka BH. Depression as a disease of modernity: explanations for increasing prevalence. J Affect Disord. 2012 Nov;140(3):205-14. doi: 10.1016/j.jad.2011.12.036

Hopcroft, R. L. (2018). Number of Childbearing Partners, Status, and the Fertility of Men and Women in the U.S Front. Sociol. 3:22. doi: 10.3389/fsoc.2018.00022

Horner, V., Whiten, A. Causal knowledge and imitation/emulation switching in chimpanzees (Pan troglodytes) and children (Homo sapiens). Anim Cogn 8, 164–181 (2005). https://doi.org/10.1007/s10071-004-0239-6

Horstman AM, Dillon EL, Urban RJ, Sheffield-Moore M. The role of androgens and estrogens on healthy aging and longevity. J Gerontol A Biol Sci Med Sci. 2012 Nov;67(11):1140-52. doi: 10.1093/gerona/gls068. Epub 2012 Mar 26. PMID: 22451474; PMCID: PMC3636678.

Huang K, Greene JD, Bazerman M. Veil-of-ignorance reasoning favors the greater good. Proc Natl Acad Sci U S A. 2019 Nov 26;116(48):23989-23995. doi: 10.1073/pnas.1910125116. Epub 2019 Nov 12. PMID: 31719198; PMCID: PMC6883824.

Jain D, Atmapoojya P, Colah R, Lodha P. Sickle Cell Disease and Pregnancy. Mediterr J Hematol Infect Dis. 2019 Jul 1;11(1):e2019040. doi: 10.4084/MJHID.2019.040. PMID: 31308916; PMCID: PMC6613624.

Jennings KJ, de Lecea L. Neural and Hormonal Control of Sexual Behavior. Endocrinology. 2020 Oct 1;161(10):bqaa150. doi: 10.1210/endocr/bqaa150. PMID: 32845294; PMCID: PMC7507403.

Kant E (1784). An Answer to the Question: What is Enlightenment? in (1996), Kant's Practical Philosophy, tr. Mary J. Gregor, Cambridge University Press, Cambridge

Kampourakis, K. Students' "teleological misconceptions" in evolution education: why the underlying design stance, not teleology per se, is the problem. Evo Edu Outreach 13, 1 (2020). https://doi.org/10.1186/s12052-019-0116-z

Kiaris, Hippokratis (2021). Genes, Polymorphisms, and the Making of Societies: A Genetic Perspective of the Divergence between East and West (Revised and Extended Edition) (2nd edition). Irvine, Boca Raton: Universal Publishers.

Kiaris H. Non-genetic linkage of personality traits and the divergence of Eastern and Western cultures: Association with Hofstede's cultural dimensions. Culture & Brain. (2022). https://doi.org/10.1007/s40167-022-00107-2.

Kiaris, H. (2022). Gene-driven Social Connectedness in the Divergence Between East and West. The Oriental Anthropologist, 22(1), 130-138. https://doi.org/10.1177/0972558X221095184

Kiaris H. The end of selection as a driver of human evolution. (Anticipations). Futures. 2022. https://doi.org/10.1016/j.futures.2022.103038.

Kiaris, Hippokratis (2023). The end of the Western Civilization? The Intellectual Journey of Humanity to Adulthood. Wilmington, DE: Vernon Press.

Kiaris H. Biology as a construct: Universals, historicity, and the postmodern critique. Perspectives in Biology and Medicine. 67 (3):337-347, 2024..

Kiaris H. Plato and political engagement. Reference to be added.

Knight, Chris. Blood Relations: Menstruation and the Origins of Culture. Yale University Press, 1991.

Krach S, Paulus FM, Bodden M, Kircher T. The rewarding nature of social interactions. Front Behav Neurosci. 2010 May 28;4:22. doi: 10.3389/fnbeh.2010.00022. PMID: 20577590; PMCID: PMC2889690.

Kraus MW, Torrez B, Park JW, Ghayebi F. Evidence for the reproduction of social class in brief speech. Proc Natl Acad Sci U S A. 2019 Nov 12;116(46):22998-23003. doi: 10.1073/pnas.1900500116. Epub 2019 Oct 21. PMID: 31636176;

Lahdenperä M, Russell AF, Tremblay M, Lummaa V. Selection on menopause in two premodern human populations: no evidence for the Mother Hypothesis. Evolution. 2011 Feb;65(2):476-89. doi: 10.1111/j.1558-5646.2010.01142.x. Epub 2010 Nov 3. PMID: 20874739.

Laredo SA, Villalon Landeros R, Trainor BC. Rapid effects of estrogens on behavior: environmental modulation and molecular mechanisms. Front Neuroendocrinol. 2014 Oct;35(4):447-58. doi: 10.1016/j.yfrne.2014.03.005. Epub 2014 Mar 29. PMID: 24685383; PMCID: PMC4175137.

Law SP. The regulation of menstrual cycle and its relationship to the moon. Acta Obstet Gynecol Scand. 1986;65(1):45-8. doi: 10.3109/00016348609158228

Laws, S. (1990). Issues of Blood: The Politics of Menstruation. London: Macmillan.

Lennox, J. G. (1993). Darwin was a teleologist. Biology and Philosophy, 8(4), 409–421.

Lennox J. Aristotle's philosophy of biology: Studies in the origins of life science

Cambridge University Press, Cambridge (2001).

Leslie M. Turner, Adrian R. Young, Holger Römpler, Torsten Schöneberg, Steven M. Phelps, Hopi E. Hoekstra, Monogamy Evolves through Multiple Mechanisms: Evidence from V1aR in Deer Mice, Molecular Biology and Evolution, Volume 27, Issue 6, June 2010, Pages 1269–1278, https://doi.org/10.1093/molbev/msq013

Lonsdorf EV, Wilson ML, Boehm E, Delaney-Soesman J, Grebey T, Murray C, Wellens K, Pusey AE. Why chimpanzees carry dead infants: an empirical assessment of existing hypotheses. R Soc Open Sci. 2020 Jul 1;7(7):200931. doi: 10.1098/rsos.200931.

Losin EA, Iacoboni M, Martin A, Dapretto M. Own-gender imitation activates the brain's reward circuitry. Soc Cogn Affect Neurosci. 2012 Oct;7(7):804-10. doi: 10.1093/scan/nsr055. Epub 2012 Mar 1. PMID: 22383803; PMCID: PMC3475355.

Lyons DE, Young AG, Keil FC. The hidden structure of overimitation. Proc Natl Acad Sci U S A. 2007 Dec 11;104(50):19751-6. doi: 10.1073/pnas.0704452104. Epub 2007 Dec 4. PMID: 18056814; PMCID: PMC2148370.

Madden S, St Pierre-Hanse N, Kelly L, Cromarty H, Linkewich B, Payne L: First nations women's knowledge of menopause: Experiences and perspectives. Candian Family Physician. 2010, 56 (9): 331-337.

Mahajan N. Fertility preservation in female cancer patients: An overview. J Hum Reprod Sci. 2015 Jan-Mar;8(1):3-13. doi: 10.4103/0974-1208.153119. PMID: 25838742; PMCID: PMC4381379.

McLanahan Sara and Percheski Christine. Family Structure and the Reproduction of Inequalities. Annual Review of Sociology 2008 34:1, 257-276

MCKUSICK VA, EGELAND JA, ELDRIDGE R, KRUSEN DE. DWARFISM IN THE AMISH I. THE ELLIS-VAN CREVELD SYNDROME. Bull Johns Hopkins Hosp. 1964 Oct;115:306-36. PMID: 14217223.

Mittnik A, Massy K, Knipper C, Wittenborn F, Friedrich R, Pfrengle S, Burri M, Carlichi-Witjes N, Deeg H, Furtwängler A, Harbeck M, von Heyking K, Kociumaka C, Kucukkalipci I, Lindauer S, Metz S, Staskiewicz A, Thiel A, Wahl J, Haak W, Pernicka E, Schiffels S, Stockhammer PW, Krause J. Kinship-based social inequality in Bronze Age Europe. Science. 2019 Nov 8;366(6466):731-734. doi: 10.1126/science.aax6219. Epub 2019 Oct 10. PMID: 31601705

Monod Jacques. Chance and Necessity—An essay on the Natural Philosophy of Modern Biology. Pp. 187. (Collins-Fontana Books, 1974.)

Mooney, T.B., Williams, J.N. (2016). The Confucian Filial Duty to Care (xiao 孝) for Elderly Parents. In: Tālivaldis Ozoliņš, J. (eds) Religion and Culture in Dialogue. Sophia Studies in Cross-cultural Philosophy of Traditions and Cultures, vol 15. Springer, Cham. https://doi.org/10.1007/978-3-319-25724-2_8

Moorad, J & Walling, C 2017, 'Measuring selection for genes that promote long life in a historical human population', Nature Ecology & Evolution, vol. 1, no. 11, pp. 1773-1781. https://doi.org/10.1038/s41559-017-0329-x

Morton RA, Stone JR, Singh RS (2013) Mate Choice and the Origin of Menopause. PLOS Computational Biology 9(6): e1003092. https://doi.org/10.1371/journal.pcbi.1003092

Muller MN, Thompson ME, Wrangham RW. Male chimpanzees prefer mating with old females. Curr Biol. 2006 Nov 21;16(22):2234-8. doi: 10.1016/j.cub.2006.09.042. PMID: 17113387

Muthukrishna M, Henrich J. Innovation in the collective brain. Philos Trans R Soc Lond B Biol Sci. 2016 Mar 19;371(1690):20150192. doi: 10.1098/rstb.2015.0192. PMID: 26926282; PMCID: PMC4780534.

Muthukrishna M, Doebeli M, Chudek M, Henrich J. The Cultural Brain Hypothesis: How culture drives brain expansion, sociality, and life history. PLoS Comput Biol. 2018 Nov 8;14(11):e1006504. doi: 10.1371/journal.pcbi.1006504

Merlan, Philip. "Aristotle's Unmoved Movers." Traditio, vol. 4, Fordham University, 1946, pp. 1–30, http://www.jstor.org/stable/27830101.

Naderi A, Soltanmaohammadi E, Kaza V, Barlow S, Chatzistamou I, Kiaris H (2021) Persistent effects of pair bonding in lung cancer cell growth in monogamous Peromyscus californicus eLife 10:e64711. https://doi.org/10.7554/eLife.64711

Nagamia, Husain F (2001). "History of Islamic Medicine". Journal of the Islamic Medical Association of North America. 28 (3). doi:10.5915/28-3-6170. ISSN 2160-9829

Obschonka M, Stuetzer M, Rentfrow PJ, Shaw-Taylor L, Satchell M, Silbereisen RK, Potter J, Gosling SD. In the shadow of coal: How large-scale industries contributed to present-day regional differences in personality and well-being. J Pers Soc Psychol. 2018 Nov;115(5):903-927. doi: 10.1037/pspp0000175.

Oliveira VEM, Bakker J. Neuroendocrine regulation of female aggression. Front Endocrinol (Lausanne). 2022 Aug 10;13:957114. doi: 10.3389/fendo.2022.957114. PMID: 36034455; PMCID: PMC9399833.

Olson R.M. (2013) Aristotle on God: Divine Nous as Unmoved Mover. In: Diller J., Kasher A. (eds) Models of God and Alternative Ultimate Realities. Springer, Dordrecht. https://doi.org/10.1007/978-94-007-5219-1_9

Oxenham, Marc F., Tilley, Lorna, Matsumura, Hirofumi, Nguyen, Lan Cuong, Nguyen, Kim Thuy, Nguyen, Kim Dung, Domett, Kate, and Huffer, Damien (2009) Paralysis and severe disability requiring intensive care in Neolithic Asia. Anthropological Science, 117 (2). pp. 107-112.

Pérusse D. Cultural and reproductive success in industrial societies: Testing the relationship at the proximate and ultimate levels. Behavioral and Brain Sciences. 1993;16(2):267-283. doi:10.1017/S0140525X00029939

Perry GH, Xue Y, Smith RS, Meyer WK, Calışkan M, Yanez-Cuna O, Lee AS, Gutiérrez-Arcelus M, Ober C, Hollox EJ, Tyler-Smith C, Lee C. Evolutionary genetics of the human Rh blood group system. Hum Genet. 2012 Jul;131(7):1205-16. doi: 10.1007/s00439-012-1147-5. Epub 2012 Feb 25. PMID: 22367406; PMCID: PMC3378649.

Piaget, J. (1971). The Theory of Stages in Cognitive Development. In D. Green, M. P. Ford, & G. B. Flamer (Eds.), Measurement and Piaget (pp. 1-11). New York, NY: McGraw-Hill.

Piel FB, Patil AP, Howes RE, Nyangiri OA, Gething PW, Williams TN, Weatherall DJ, Hay SI. Global distribution of the sickle cell gene and geographical confirmation of the malaria hypothesis. Nat Commun. 2010 Nov 2;1:104. doi: 10.1038/ncomms1104. PMID: 21045822; PMCID: PMC3060623.

Pomeroy E, Bennett P, Hunt CO, et al. New Neanderthal remains associated with the 'flower burial' at Shanidar Cave. Antiquity. 2020;94(373):11-26. doi:10.15184/aqy.2019.207

Proctor D, Brosnan SF, de Waal FB. How fairly do chimpanzees play the ultimatum game? Commun Integr Biol. 2013a May 1;6(3):e23819. doi: 10.4161/cib.23819. PMID: 23713135; PMCID: PMC3656016.

Proctor, D., Williamson, R. A., de Waal, F. B. M., & Brosnan, S. F. (2013b). Chimpanzees play the ultimatum game. PNAS Proceedings of the National Academy of Sciences of the United States of America, 110(6), 2070–2075. https://doi.org/10.1073/pnas.1220806110.

Rafferty, John P. "biodiversity loss". Encyclopedia Britannica, 23 Dec. 2023, https://www.britannica.com/science/biodiversity-loss.

Reich, Robert B., The Common Good. New York, Alfred A. Knopf, 2018.

Ramani PK, Parayil Sankaran B. Tay-Sachs Disease. [Updated 2023 Jan 25]. In: StatPearls [Internet]. Treasure Island (FL): StatPearls Publishing; 2023 Jan-. Available from: https://www.ncbi.nlm.nih.gov/books/NBK564432/

Rawls, J. (1971). A Theory of Justice: Original Edition. Harvard University Press. https://doi.org/10.2307/j.ctvjf9z6v

Rejón, Manuel Ruiz. The Origin of Species in the Anthropocene. 2021. OpenMind: BBVA's knowledge community. https://www.bbvaopenmind.com/en/science/bioscience/origin-of-species-in-the-anthropocene.

Rendu W, Beauval C, Crevecoeur I, Bayle P, Balzeau A, Bismuth T, Bourguignon L, Delfour G, Faivre JP, Lacrampe-Cuyaubère F, Tavormina C, Todisco D, Turq A, Maureille B. Evidence supporting an intentional Neandertal burial at La Chapelle-aux-Saints. Proc Natl Acad Sci U S A. 2014 Jan 7;111(1):81-6. doi: 10.1073/pnas.1316780110

Ridley Matt. The Origins of Virtue: Human Instincts and the Evolution of Cooperation. 1998 Penguin Publishing Group. ISBN 9780140264456

Rivollat M, Thomas A, Ghesquière E, Rohrlach AB, Späth E, Pemonge MH, Haak W, Chambon P, Deguilloux MF. Ancient DNA gives new insights into a Norman Neolithic monumental cemetery dedicated to male elites. Proc Natl Acad Sci U S A. 2022 May 3;119(18):e2120786119. doi: 10.1073/pnas.2120786119. Epub 2022 Apr 21. PMID: 35446690; PMCID: PMC9170172

Ross CT, Hooper PL, Smith JE, Jaeggi AV, Smith EA, Gavrilets S, Zohora FT, Ziker J, Xygalatas D, Wroblewski EE, Wood B, Winterhalder B, Willführ KP, Willard AK, Walker K, von Rueden C, Voland E, Valeggia C, Vaitla B, Urlacher S, Towner M, Sum CY, Sugiyama LS, Strier KB, Starkweather K, Major-Smith D, Shenk M, Sear R, Seabright E, Schacht R, Scelza B,

Roth, Martha Tobi & Hoffner, Harry A & Michalowski, Piotr. (1997). Law collections from Mesopotamia and Asia Minor / by Martha T. Roth ; with a contribution by Harry A. Hoffner, Jr. ; volume editor, Piotr Michalowski. Atlanta, Ga. : Scholars Press

Rohrlach, A.B., Rivollat, M., de-Miguel-Ibáñez, P. et al. Cases of trisomy 21 and trisomy 18 among historic and prehistoric individuals discovered from ancient DNA. Nat Commun 15, 1294 (2024). https://doi.org/10.1038/s41467-024-45438-1

Romm Sharom. Beauty thtough history. The changing ideal. The Washington Post. January 26, 1987

Sartre, Jean-Paul. Existentialism Is a Humanism. Translated by Carol Macomber, Yale University Press, 2007.

Schimmelpfennig R, Razek L, Schnell E, Muthukrishna M. Paradox of diversity in the collective brain. Philos Trans R Soc Lond B Biol Sci. 2022 Jan 31;377(1843):20200316. doi: 10.1098/rstb.2020.0316. Epub 2021 Dec 13. PMID: 34894736

Scott TW. Crozier's paradox and kin recognition: Insights from simplified models. J Theor Biol. 2024 Mar 21;581:111735. doi: 10.1016/j.jtbi.2024.111735. Epub 2024 Jan 19. PMID: 38246487.

Skirbekk V. 2008 Fertility trends by social status. Demogr. Res. 18, 145–180. (doi:10.4054/DemRes.2008.18.5).

Slaby RG, Frey KS. Development of gender constancy and selective attention to same-sex models. Child Development. 1975;46:849–56.

Scaggs S, Salerno J, Revilla-Minaya C, Redhead D, Pusey A, Purzycki BG, Power EA, Pisor A, Pettay J, Perry S, Page AE, Pacheco-Cobos L, Oths K, Oh SY, Nolin D, Nettle D, Moya C, Migliano AB, Mertens KJ, McNamara RA, McElreath R, Mattison S, Massengill E, Marlowe F, Madimenos F, Macfarlan S, Lummaa V, Lizarralde R, Liu R, Liebert MA, Lew-Levy S, Leslie P, Lanning J, Kramer K, Koster J, Kaplan HS, Jamsranjav B, Hurtado AM, Hill K, Hewlett B, Helle S, Headland T, Headland J, Gurven M, Grimalda G, Greaves R, Golden CD, Godoy I, Gibson M, Mouden CE, Dyble M, Draper P, Downey S, DeMarco AL, Davis HE, Crabtree S, Cortez C, Colleran H, Cohen E, Clark G, Clark J, Caudell MA, Carminito CE, Bunce J, Boyette A, Bowles S, Blumenfield T, Beheim B, Beckerman S, Atkinson Q, Apicella C, Alam N, Mulder MB. Reproductive inequality in humans and other mammals. Proc Natl Acad Sci U S A. 2023 May 30;120(22):e2220124120. doi: 10.1073/pnas.2220124120. Epub 2023 May 22. PMID: 37216525.

Schoenaker DA, Jackson CA, Rowlands JV, Mishra GD. Socioeconomic position, lifestyle factors and age at natural menopause: a systematic review and meta-analyses of studies across six continents. Int J Epidemiol. 2014 Oct;43(5):1542-62. doi: 10.1093/ije/dyu094. Epub 2014 Apr 26. PMID: 24771324; PMCID: PMC4190515.

Shenk MK, Kaplan HS, Hooper PL. Status competition, inequality, and fertility: implications for the demographic transition. Philos Trans R Soc Lond B Biol Sci. 2016 Apr 19;371(1692):20150150. doi: 10.1098/rstb.2015.0150

Smith, ST, and Buzon, MR. 'Cross-Cultural Interaction in the Ancient Egyptian and Nubian Borderland', in Ulrike Matthies Green, and Kirk E. Costion (eds), Modeling Cross-Cultural Interaction in Ancient Borderlands (Gainesville, FL, 2018; online edn, Florida Scholarship Online, 20 Sept. 2018), https://doi.org/10.5744/florida/9780813056883.003.0005,

Solinas, Marco (2015). From Aristotle's Teleology to Darwin's Genealogy: The Stamp of Inutility, Basingstoke: Palgrave Macmillan, 2015 (pdf: Contents, Introduction). Palgrave-Macmillan.

Song X, Massey CG, Rolf KA, Ferrie JP, Rothbaum JL, Xie Y. Long-term decline in intergenerational mobility in the United States since the 1850s. Proc Natl

Acad Sci U S A. 2020 Jan 7;117(1):251-258. doi: 10.1073/pnas.1905094116. Epub 2019 Nov 25. PMID: 31767745

Southerton, D. (2011). Mass culture (frankfurt school). In Encyclopedia of Consumer Culture (Vol. 3, pp. 904-907). SAGE Publications, Inc., https://doi.org/10.4135/9781412994248

Sowers MR, Jannausch ML, McConnell DS, Kardia SR, Randolph JF Jr. Endogenous estradiol and its association with estrogen receptor gene polymorphisms. Am J Med. 2006 Sep;119(9 Suppl 1):S16-22. doi: 10.1016/j.amjmed.2006.07.002. PMID: 16949384.

Spengler, Oswald. The Decline of the West. V. 1, Alfred A. Knopf, 1926, pp. 46–47.

Spikins Penny, Needham Andy, Tilley Lorna & Hitchens Gail (2018) Calculated or caring? Neanderthal healthcare in social context, World Archaeology, 50:3, 384-403, DOI: 10.1080/00438243.2018.1433060

Spiteri T, Musatov S, Ogawa S, Ribeiro A, Pfaff DW, Agmo A. The role of the estrogen receptor alpha in the medial amygdala and ventromedial nucleus of the hypothalamus in social recognition, anxiety and aggression. Behav Brain Res. 2010 Jul 11;210(2):211-20. doi: 10.1016/j.bbr.2010.02.033. Epub 2010 Feb 23. PMID: 20184922.

Stone Lyman (2020). The Rise of Childless America. Institute for family studies. https://ifstudies.org/blog/the-rise-of-childless-america

Strassmann JE, Gilbert OM, Queller DC. Kin discrimination and cooperation in microbes. Annu Rev Microbiol. 2011;65:349-67. doi: 10.1146/annurev.micro.112408.134109. PMID: 21682642.

Takahashi M, Singh RS, Stone J. A Theory for the Origin of Human Menopause. Front Genet. 2017 Jan 6;7:222. doi: 10.3389/fgene.2016.00222. PMID: 28111590; PMCID: PMC5216033.

Tennie C, Call J, Tomasello M. 2009. Ratcheting up the ratchet: on the evolution of cumulative culture. Philos. Trans. R. Soc. B 364(1528): 2405–15

Scheubel, Beatrice. "Bismarck's Pension System." Bismarck's Institutions: A Historical Perspective on the Social Security Hypothesis, Mohr Siebeck GmbH and Co. KG, 2013, pp. 77–104. JSTOR, http://www.jstor.org/stable/j.ctvf3w2sm.8. Accessed 10 Mar. 2024.

Thapa, S., Aro, A.R. 'Menstruation means impurity': multilevel interventions are needed to break the menstrual taboo in Nepal. BMC Women's Health 21, 84 (2021). https://doi.org/10.1186/s12905-021-01231-6.

Tilley, L. 2012. "The Bioarchaeology of Care." The SAA Archaeological Record 12 (3): 39–41.

Tilley, L., and T. Cameron. 2014. "Introducing the Index of Care: A Web-Based Application Supporting Archaeological Research Into Health-Related Care." International Journal of Paleopathology 6: 5–9.

Tilley L, Oxenham MF. Survival against the odds: Modeling the social implications of care provision to seriously disabled individuals. Int J Paleopathol. 2011 Mar;1(1):35-42. doi: 10.1016/j.ijpp.2011.02.003. Epub 2011 Mar 29. PMID: 29539340.

Tomasello, M Carpenter, J Call, T Behne, H Moll, Understanding and sharing intentions: The origins of cultural cognition. Behav Brain Sci 28, 675–691, discussion 691–735 (2005).

Tomasello M, Great apes and human development: A personal history. Child Dev Perspect, in press. (February 10, 2018).

Tulchinsky TH. Bismarck and the Long Road to Universal Health Coverage. Case Studies in Public Health. 2018:131–79. doi: 10.1016/B978-0-12-804571-8.00031-7. Epub 2018 Mar 30. PMCID: PMC7149836.

Vining DR. Social versus reproductive success: The central theoretical problem of human sociobiology. Behavioral and Brain Sciences. 1986;9(1):167-187. doi:10.1017/S0140525X00021968

Voegeli R, Schoop R, Prestat-Marquis E, Rawlings AV, Shackelford TK, Fink B. Cross-cultural perception of female facial appearance: A multi-ethnic and multi-centre study. PLoS One. 2021 Jan 22;16(1):e0245998. doi: 10.1371/journal.pone.0245998. PMID: 33481957

Von Rueden, Christopher R., and Adrian V. Jaeggi. 2016. "Men's Status and Reproductive Success in 33 Nonindustrial Societies: Effects of Subsistence, Marriage System, and Reproductive Strategy." Proceedings of the National Academy of Sciences 113 (39). Washington, D.C: National Academy of Sciences: 10824–29.

Wang, D.Y., Key, T.J.A., Pike, M.C. et al. Serum hormone levels in British and rural Chinese females. Breast Cancer Res Tr 18 (Suppl 1), S41–S45 (1991). https://doi.org/10.1007/BF02633526

Weyrich, L., Duchene, S., Soubrier, J. et al. Neanderthal behaviour, diet, and disease inferred from ancient DNA in dental calculus. Nature 544, 357–361 (2017). https://doi.org/10.1038/nature21674

Wharton W, Gleason CE, Olson SR, Carlsson CM, Asthana S. Neurobiological Underpinnings of the Estrogen - Mood Relationship. Curr Psychiatry Rev. 2012 Aug 1;8(3):247-256. doi: 10.2174/157340012800792957. PMID: 23990808; PMCID: PMC3753111.

White, Michael J. "Unmoved Movers, Celestial Spheres, and Cosmoi: Aristotle's Diremption of the Divine" Apeiron, vol. 55, no. 1, 2022, pp. 97-118. https://doi.org/10.1515/apeiron-2020-0063

Whiten A, Ayala FJ, Feldman MW, Laland KN. The extension of biology through culture. Proc Natl Acad Sci U S A. 2017 Jul 25;114(30):7775-7781. doi: 10.1073/pnas.1707630114. Epub 2017 Jul 24. PMID: 28739924; PMCID: PMC5544333.

Whiten A. Social Learning and Culture in Child and Chimpanzee. Annual Review of Psychology 2017 68:1, 129-154

Wilson, E.O.. "Mass Extinction". Encyclopedia Britannica, 3 Aug. 2023, https://www.britannica.com/topic/Mass-Extinction-2117803.

Wilson, E. O., 1975, Sociobiology, Cambridge, MA: Belknap/Harvard University Press.

Wilson EO Hölldobler B Eusociality: Origins and consequences Proceedings of the National Academy of Sciences 2005 102 13367 13371

Xu L, Au Yeung SL, Kavikondala S, Leung GM, Schooling CM. Testosterone concentrations in young healthy US versus Chinese men. Am J Hum Biol. 2014 Jan-Feb;26(1):99-102. doi: 10.1002/ajhb.22482. Epub 2013 Nov 19. PMID: 24254402.

Xu L, Yeung SL, Kavikondala S, Schooling CM. Estradiol concentrations in young healthy US versus Chinese men. Am J Hum Biol. 2014 Jul-Aug;26(4):565-9. doi: 10.1002/ajhb.22548. Epub 2014 Apr 3. PMID: 24700607.

Zerjal T, Xue Y, Bertorelle G, Wells RS, Bao W, Zhu S, Qamar R, Ayub Q, Mohyuddin A, Fu S, Li P, Yuldasheva N, Ruzibakiev R, Xu J, Shu Q, Du R, Yang H, Hurles ME, Robinson E, Gerelsaikhan T, Dashnyam B, Mehdi SQ, Tyler-Smith C. The genetic legacy of the Mongols. Am J Hum Genet. 2003 Mar;72(3):717-21. doi: 10.1086/367774. Epub 2003 Jan 17. PMID: 12592608; PMCID: PMC1180246.

Made in United States
North Haven, CT
04 November 2026

21760273R00118